There's Nothing There

but

Nothing is Really Something

Andrew Fletcher

There's Nothing There

but

Nothing is Really Something

Andrew Fletcher

Copyright © 2018 by Andy Fletcher

All rights reserved. No part of this book may be reproduced in any form or by any electronic or mechanical means, including information storage and retrieval systems, without permission in writing from the publisher, except by a reviewer, who may quote brief passages in a review.

Published by fletchpub at Lulu

Additional copies found at www.lulu.com/spotlight/andyfletch.

ISBN: 978-1-387-63721-8

Set in 12 point Garamond type

First Edition 2018

Front Cover Photo: "Van Gogh from Space", Gotland Island, Sweden, by Landsat, NASA Goddard Space Flight Center

Back Cover Photo: "Ira's Ghost", Hubble Space Telescope

The science in this book is used in a series of seminars developed and given worldwide by LUE, Inc. For information on hosting the seminars, to see the bibliography from which the seminars were developed, to see a list of places where the seminars have been given, to buy the seminars in book or DVD form, or for a history of the seminars and the organization, please visit www.lifeuniverseverything.org.

Life, the Universe, and Everything, Inc. Also known as TOK Seminars.
c/o TOK Seminars,
PO Box 62627
Colorado Springs CO 80920 USA

+1.719.660.2602

For information on hosting the talks in your school or other venue, go to http://andyfletch3.wix.com/fletchblog.

Table of Contents

Page	Chapter	Title
11	1	What a Great Universe
13	2	There's Nothing There Part 1
17	3	Loving Big Bang
23	4	There's Nothing There Part 2
25	5	There's Nothing There Part 3
29	6	There's Nothing There Part 4
33	7	There's Nothing There Part 5
35	8	There's Nothing There Part 6
41	9	There's Nothing There Part 7
47	10	There's Nothing There Part 8
51	11	There's Nothing There Part 9
55	12	There's Nothing There Part 10
59	13	There's Nothing There Part 11
63	14	There's Nothing There Part 12
67	15	There's Nothing There Part 13
71	16	There's Nothing There Part 14
75	17	There's Nothing There Part 15
79	18	There's Nothing There Part 16
85	19	There's Nothing There Part 17
91	20	There's Nothing There Part 18
97	21	There's Nothing There Part 19
101	22	There's Nothing There Part 20
107	23	There's Nothing There Part 21
111	24	There's Nothing There Part 22
117	25	There's Nothing There Part 23
121	26	There's Nothing There The End
125	27	There's Something There - Part 1?

Page	Chapter	Title
131	28	There's Something There - Part 2
137	29	There's Something There - Part 3
141	30	There's Something There - Part 4
145	31	There's Something There - Part 5
149	32	There's Something There - Part 6
153	33	There's Something There - Part 7
157	34	Gaps
163	35	Gap are Back
167	36	Free Will or Free Won't
171	37	A Brief History of Free Will
175	38	Free Will is Back. Sort of.
181	39	Free Will and Butterflies
185	40	Free Will and Harley Guys
189	41	Free Will. Get Over It. Or Not.
195	42	Now It Gets Confusing.
199	43	No Wonderful Plan for Your Life
203	44	Bad Things, Good People. Chaos.
207	45	Life is Chaotic. Then You Die.
211	46	Living Within the Clumps
215	47	Something Else Going On
219	48	Talkin' About the E-word
225	49	Evolving Evolution
229	50	Not Your Grandmothers Evolution
235	51	What's It All About?
241	52	The Talking Bacteria
245	53	Community
249	54	All You Need is Love

Introduction

There's nothing there.

Kind of a wild claim. I'm not going to give it away here in the introduction. Then you wouldn't have to read the book.

It does create the immediate possibility that book reviewers will use it in what they think is clever writing to mount a critique of this book.

As in, there's nothing there in There's Nothing There.

Ha. Hysterical. Plus, now you have to come up with something else to write. I'm sure you'll manage. Allow me to suggest something along the lines of there's something there in There's Nothing There. And it's really something. That would be lovely.

Anyway.

It seemed like a reasonable assumption for everyone to make that there was actually something there.

I mean, you could see it.

The sun. The moon. The planets. Their moons. Other stars. The Milky Way. Eventually, other galaxies. Lots and lots of galaxies. Gazillions.

And back here on Earth, all kinds of things. Flowers. Rainbows. Laughing children. Young love. Plus, wars and genocides and tsunamis and lava flow and zits. Reality shows. The Kardashians.

There was clearly something there.

Turns out, not so much.

Everything you think is there is not really there the way that you think it should be.

Do all the scientists agree with that observation?

Turns out, pretty much.

So what do all the scientists think is actually there?

Now that's where it gets interesting. You'll enjoy this.

This was all originally written as a blog, which in turn derived from the talks we give on Life, the Universe and Everything to people all over the world. But there are two other books that led to this one, so you should go buy them, too.

But this is new. Frankly, it's just very cool. You should read it. Americans would say that it's awesome. Brits would say, I don't know, spot of tea or bangers and mash or something.

All of your friends and relatives would say, you should have bought them a copy.

I'd go do that right now.

Chapter 1   What a Great Universe

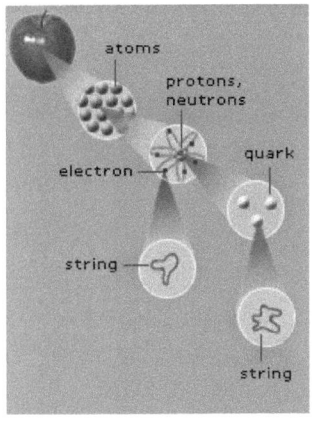

Just to open with an editorial comment on the marvelous state of our understanding of the universe:

1. It's been ~40+ years since String Theory was proposed, and there's still not a single thing we know to be true about it. There is not one String Theory fact.

2. It's been ~30+ years since Inflationary Theory was proposed, and there's still no evidence in support of it. Almost, but still nothing definitive.

3. It's been ~40+ years since we discovered the need for Dark Matter to exist, and we still haven't a clue what it is.

4. It's been 15+ years since we discovered a need for Dark Energy to exist, and we still haven't a clue what it is.

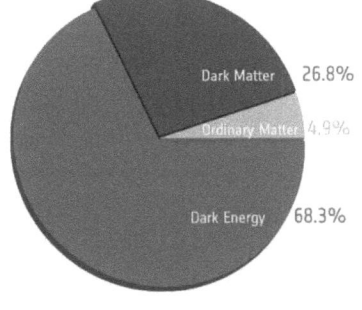

4.5 Dark Energy is 68% of what the universe is made of. There's 5-6 times more Dark Matter than regular matter, so Dark Matter is 27% of what the universe is made of. Thus 95% of the universe is a complete and total mystery to us. Maybe even 96%.

5. It's been ~115+ years since Quantum Theory appeared on the scene to explain how the universe is put together at the level of the very small, and it explains it perfectly, and there are legions of facts and supporting evidence, and it's never wrong and it's always right, and it still makes no sense to anybody and nobody likes it and everyone wishes it would just go away. And it's the finest scientific theory known to man.

6. It's been ~110+ years since Special and General Relativity arrived on the scene to explain how the universe is put together at the level of the very large, and they explain it brilliantly and they're always right and never wrong. And they still don't make any sense.

6.5 General Relativity led us to Black Holes and Big Bang. There are legions of facts and supporting evidence for Big Bang, and there's no evidence against it, and most scientists who really understand it are not happy with it and wish it would just go away, and most people of faith refuse to believe in it because they don't really understand it. And Black Holes don't exist. They almost exist and they'll be at the edge of existing forever, or until they evaporate, which apparently they will all do, so actually, Black Holes don't quite exist and never will.

7. It's been ~110+ and ~115+ years since General Relativity and Quantum Theory arrived on the scene to explain the way the universe works in its entirety, and they're both perfect and they're never wrong, and they're the two finest fields of science known to man, and they are still entirely incompatible and don't get along with each other.

What a great universe.

Chapter 2   There's Nothing There – Part 1

We tend to think that there's something there. That is, when you look around, you see stuff, and we assume that the stuff is actually made of stuff. Well.

The numbers below are mostly and necessarily approximate, and they change all the time, and I had to figure some of them out, and I don't guarantee my figurin', and I left out the interstellar gases. But it'll do.

(PS I've been told that every time you use a bullet point, a kitten dies. So for no logical reason, I used the infinity symbol.)(Cats are evil. It's in my other book.)

The Universe
- ∞ The observable universe is ~93 billion light years across and has ~1 trillion galaxies. Maybe 2 trillion. Or 20. We think.
- ∞ There are more stars in the observable universe than grains of sand on earth. Maybe 10 times as much. Maybe more.
- ∞ The Milky Way contains 300 billion stars. We think. Maybe 400 billion.
- ∞ It's 100,000 light years in diameter and 1000 light years thick.
- ∞ That's (600,000,000,000,000,000) 600 million billion miles in diameter, and 6 million billion (6,000,000,000,000) miles thick.
- ∞ The nearest star to us is 4+ light years away, 24 trillion miles of empty space - 24,000,000,000,000 miles of nothing between us and it.
- ∞ "There is only one star in every 100 billion cubic light-years of space and the average distance between stars in the universe is ~ 4,150 light-years, or about 100 times the distance between the Sun and its nearest stellar neighbor" (Astronomy Mag., Mar 13, 2014)
- ∞ 1 light year^3 = 2×10^{38} cubic miles
- ∞ That's 1 star in every 20 trillion trillion trillion trillion cubic miles

- ∞ Star density in a globular cluster (right there in that picture)(very dense) is about 1 star per every 800 billion trillion trillion cubic miles of space.
- ∞ Star density in the Milky Way - about 1 star per every 3 thousand trillion trillion trillion cubic miles of space.

Therefore. Space is empty. There's nothing there.

- ∞ Our solar system has 8 planets
- ∞ It's 2.8 billion miles to Neptune
- ∞ That's 8 planets spread over 2.8 billion miles
- ∞ That's 1 sun and 8 planets in about 91 billion billion billion cubic miles of space.

Therefore. The Solar System is empty. There's nothing there.

Matter (the solid stuff)

- ∞ There's about 10^{50} tons of matter in the universe.
- ∞ There's about 1 hydrogen atom per 3 cubic meters of space on average.
- ∞ Nobody's ever seen an atom.

Therefore. There's nothing there.

Atoms (what matter is made of) and You

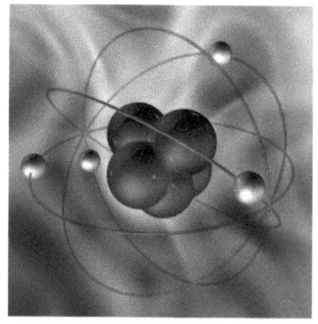

- ∞ What is an atom? Protons, neutrons, electrons, and emptiness.
- ∞ If a proton is blown up to ~1000 pixels (about the size of a large grapefruit), the nearest electron would be 11 miles away
- ∞ There's a million billion times more empty space than matter in an atom. (1,000,000,000,000,000)

- ∞ You are made up of ~70 trillion cells. Each cell has 100 trillion atoms. That's 7 thousand trillion trillion atoms in you. Roughly. (7,000,000,000,000,000,000,000,000,000)
- ∞ There's a million billion times more nothing than something in you. You are almost entirely empty space.
- ∞ If you squeeze out all the empty space from all the atoms from all the 7.5 billion people on the planet, everyone fits into an M&M. A Smarty, for you Europeans.
- ∞ Actually, it's 2/3s of an M&M. There's room for another 3 billion people.

Therefore. There's nothing there.

Protons etc.

- ∞ Protons are made of 3 quarks
- ∞ Each quark is 10^7 times smaller than the atom.
- ∞ 3 quarks make up only 1% of the mass of the proton.
  - ∞ The rest?
  - ∞ Virtual particles that flit in and out of existence, particles and anti-particles that instantly annihilate each other. It's $e=mc^2$ all the time.
  - ∞ 99% of your mass is virtual.
  - ∞ 99% of the you that is actually there, is only temporarily there in a constant sort of way.

Therefore. There's nothing there.

Particles etc.

- ∞ Particles are just energy anyway.
- ∞ Particles have no real position in space or time - they are just probability waves.
- ∞ Protons and neutrons are made of quarks.

- ∞ Quarks have no real existence.
- ∞ Everything solid is made of things that don't really exist.
- ∞ "Everything we call real is made up of things that cannot be regarded as real." Neils Bohr, the father of Quantum Theory, the theory of what matter actually is.

Therefore. There's nothing there.

Back to the Universe

- ∞ There's really just nothing in the universe but an occasional bit of something that's not really there made out of energy that becomes matter.
- ∞ How much energy is in the universe?
- ∞ None. There's no net energy in the universe.
- ∞ At Big Bang, the universe divided zero energy into two types of energy - positive and negative. We think. We're not sure. But if it's true, then …
- ∞ Positive energy became matter.
- ∞ Negative energy became gravity.
- ∞ They exactly cancel each other out.
- ∞ There's no net energy in the universe.

Therefore. There's nothing there.

That would be really depressing if it were true. OK, it's true. But it's not depressing. Watch this space. Actually, the next space.

Chapter 3   Why Religious People should love Big Bang. Not the show. The Bang.

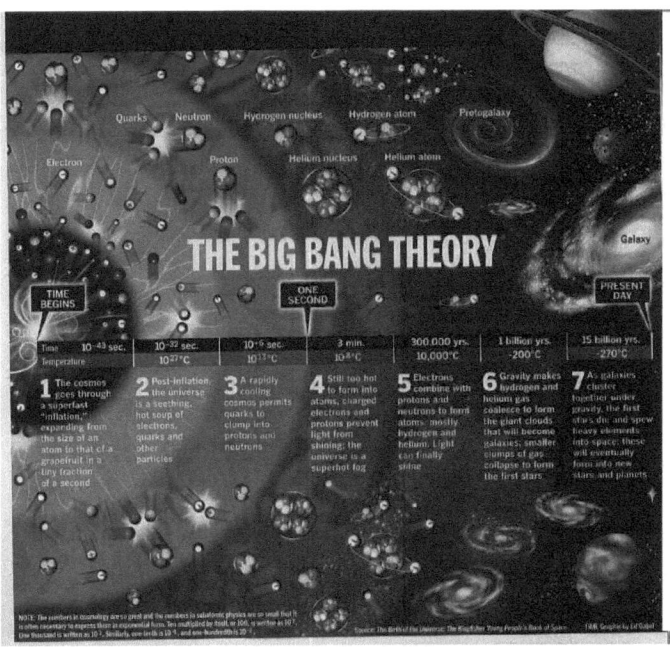

Let's just start with the answer. You should love Big Bang because the discovery of Big Bang was the most amazing, stunning, unbelievable, extraordinary, mind-boggling, gob-smacking, paradigm-shattering, science-altering event in the history of human thought, and it brought the discussion over the existence of God back onto the table.

You should love Big Bang because Albert Einstein rejected it out of hand as ridiculous, and messed up a perfectly good equation trying to force the universe to, well, not have a Big Bang right there at the beginning. Not only did Einstein reject it, most if not all scientists rejected it. Einstein rejected it even though his own personal General Theory of Relativity predicted it a full decade before there was any evidence that it might be true. This was a big surprise to Albert. He didn't put a Big Bang in there, so it was quite a shock when it popped out. It was really shocking to him when all the evidence started to line up in support of Big Bang. He had to go back and fix his equation, which made it beautiful once again.

You should love Big Bang because British astronomer Fred Hoyle rejected it to the end of his life in 2001, saying in 1982 that "the passionate frenzy with which the Big Bang cosmology is clutched to the corporate scientific bosom evidently arises from a deep-rooted attachment to the first chapter of Genesis, religious fundamentalism at its strongest." He rejected it because it sounded too much like religion.

And many reject it because it sounds too much like science.

Ah, the irony.

You especially should love Big Bang because if you don't, if you try to argue for a young earth and universe that are only 6000 years old, not only do skeptics and atheists dismiss that argument out of hand, they think you're an idiot, and you've lost any chance you might have had to talk about the real and serious issues of faith.

Here's the thing - I'm guessing that most Protestant believers reject Big Bang in large part for the same reason that most science-minded skeptics accept it without question. That reason is pretty simple - neither group has the slightest understanding of what Big Bang really is.

So it might be useful to explain it. 1-2-3 go.

We all understand the universe to be the way that Isaac Newton's equations describe it for us. It's logical, reasonable. It makes sense. It is commonsensical. It's a WYSIWYG universe - what you see is what you get. When you look at it, that's the way it is.

Yeah. Not so much.

Scientists had decided that since the universe looked really big and really old, that it must in fact be infinitely big and infinitely old. It had always been there, static, stationary, unchanging. That's what it looked like. That must be the way it is, was, and had been.

Turns out not to be that way. At all.

In 1905 and 1915, Einstein turned our understanding of the universe upside down and inside out. He even turned his own understanding of the universe on its head.

In 1905 in the Special Theory of Relativity, he discovered that time is not a constant. It's a variable. It changes as you get closer to the speed of light. The closer you get, the more it changes. Time has a speed, and its speed slows down as you speed up. The faster you go, the slower time passes.

And if you could reach the speed of light, time would stop altogether. Actually, it's more accurate to say that at the speed of light, things that happen aren't separated by time anymore. Everything happens at once. All of the history of the universe happens at the same instant. Your birth, your life, your death, everything happens at the same time.

What's more, he discovered that space and time aren't separate - they're woven together. The universe is made of space-time. And at the speed of light, space flattens to two dimensions. So not only does everything happen at once, it all happens at the same place. Sort of. I have this physics t-shirt that reads: Time exists so that everything doesn't happen at once. Space exists so that it all doesn't happen to you. Ha. Hysterical.

In 1915 in the General Theory of Relativity, he revealed his discovery that space-time is flexible, bendable, warpable. And what bends and warps space-time is matter. Anything made of matter bends and warps space-time. What we call that is Gravity. Gravity is the bending of space-time by anything made of matter. The sun, the earth, the moon. You. Me. Matter of any size and shape. We all bend space-time.

Among other things, this means that the closer you are to center of the earth, for example, the slower time passes. Time passes more slowly at your toes than at your nose.

Among other other things, this means that the more matter you have squished into a very small place, the greater the warping of space-time. Space-time can warp so much that it bends completely around itself into a sphere and cuts off the universe. We call that a Black Hole.

Now it gets interesting. Like, if it wasn't already.

Newton's fixed space   Einstein's flexible space-time

A guy named Georges Lemaître, a Belgian physicist who was also a priest, worked through the math of the General Theory in 1927 and discovered something that Albert himself didn't put there, didn't expect, and didn't want. Lemaître found out that the General Theory predicts that the universe will be expanding as you go forward in time. The universe expands. It doesn't stay the same size. It's not static, stationary, or unchanging.

And if you turn around and look backwards in time, then the universe is getting smaller. It's contracting. And it can only contract just ... so ... far ... before it ... disappears altogether.

Fred Hoyle, in trying to explain this on the radio in 1949, called it a "Big Bang." The name stuck. He didn't like it. Nobody else did at first, either. Not the name. The name was fine. The concept. The universe having a starting point.

It was just something the General Theory predicted. There was no evidence for it. Without evidence, it's not science yet, neither true nor false.

Then.

In 1929, Edwin Hubble revealed his discovery that in looking at the light of the distant galaxies in the universe, he found that the universe was in fact expanding. It was getting bigger. Lemaître was right, Einstein was wrong. There was a starting point, a place where it all began.

Now.

What does THAT mean?

Here's what we tend to think. We tend to think that there was a vast emptiness, dark, an endless vacuum of nothingness, no planets, no stars, no galaxies, no Walmarts, no Starbucks. Nothing but nothing.

That's wrong. It implies that nothing is, in fact, something. That something would be space-time.

But as you go backwards in time, the universe contracts. Space-time contracts. Space-time itself gets smaller and smaller until it just … disappears altogether.

Which means this: Big Bang was that moment when space and time came into existence.

Before Big Bang, there was no … before. No space. No time. No nothing. There was no there there for anything to be there in, no when then for anything to be then in.

Then, in a tiny tiny fraction of a second, the rate of expansion perfect to one part in ten to the sixtieth, the universe blew itself into existence. In a moment of a moment of a moment, what we call the Singularity became a universe-sized universe, a cosmos-sized cosmos, the whole process taking far less than one second.

Science thinks that one part of the Big Bang was when the universe expanded from a nano-sized square, the size of a proton, to 250 million light years across in .0000000000000000000000000000000001 second.

Everything … came from … nothing. In, like, no time.

Space. Time. Energy. Matter. The laws of physics themselves.

Everything and the potential for everything else came from … nothing.

I told this to a classroom full of students in Switzerland a few years ago. There was a long, long, quiet pause when nobody breathed and everyone tried to absorb it. They were mostly skeptics, atheists who believed that science has disproved the existence of God, that Big Bang was central to showing that God does not exist.

Then I told them about the Bang. Quiet. And then someone in the back of the room, some atheist, skeptical kid, said one word. He said, "God."

God.

That's why you should love the Big Bang.

Chapter 4   There's Nothing There – Part 2

So. According to one way of looking at the universe, there's really nothing in it. And that's pretty much right. At the smallest level of "something", on the level of electrons, quarks and gluons (which make up protons and neutrons and atoms and hence all the matter in the cosmos), electrons, quarks and gluons aren't
really there. It's a Quantum thing. And as far as everything bigger than little things goes, it's all mostly empty space, the parts that aren't empty being occupied by things (matter) made up of things (smaller matter) that aren't really there.

Clearly, that needs fixin'.

OK, then. Here we go.

First, Chaos Theory. Chaos Theory says that the universe is radically unpredictable (think "Jurassic Park"), and the reason that it's unpredictable is because there are too many tiny little things that we can't know about that make bigger things happen. The Butterfly Effect. A butterfly flaps its wings off the coast of South American and you get a hurricane (instead of not a hurricane) in the Atlantic.

Like, a fruit vendor in Tunisia gets hacked off at his government, sets himself on fire, the dominoes that became Arab Spring start falling over, and still nobody knows what the end result will be.

Or, a tiny rubber ring on the space shuttle gets a bit cold, loses a bit of its elasticity, and the Challenger blows up 72 seconds into its launch. Or a 1 kg bit of insulating foam hits the wing in the wrong way, makes a tiny little crack, and the Columbia disintegrates on reentry into the earth's atmosphere.

One tree branch falls on one power line in Ohio and causes the (then) largest blackout in North American history, hundreds of miles away in New York and eastern Canada. Another falls on a power line in Switzerland and blacks out all of ... Italy.

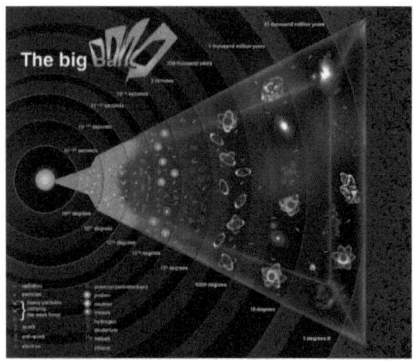

Tiny things can and do have a huge impact on the way universe does its business.

So, second. Big Bang.

Big Bang came from a Singularity. This Singularity was an infinitely compressed point of pure energy potential.

It had no size. It had no substance. Nothing was in it. Not matter, not energy. Nothing. It was the tiniest of all possible things, since it wasn't even there in any sense that we understand about "being there", especially since there was no there for it to be there in. Time and space didn't exist yet. The potential for them was in the Singularity.

And in the tiniest of all possible times, the universe blew itself into existence. As we said a minute ago, one theory suggests that in a trillionth of a trillionth of a trillionth of a second, the universe went from a nanometer square to 250 million light years across. Our own Milky Way is a puny 100,000 light years across, an embarrassingly small 1000 light years thick.

In .0000000000000000000000000000000001 of a second.

1,500,000,000,000,000,000,000 miles across.

Everything coming from nothing in as close to no time as it's possible to imagine.

Time and space and energy and matter and the potential for everything that became everything that is. From nothing.

Tiny things seem to be able to do very, very large things.

Now we have a cool question to deal with.

Since the universe arrived, that is, since it used to not be here, and now it is, where did everything inside the universe come from? Why did the universe do anything? Why is there something instead of nothing?

Technically, that's three questions. *Tant pis*, as they say in France.

Chapter 5   There's Nothing There – Part 3

So here's what we resolved - everything came from nothing. So nothing actually matters. That is, the nothingness that gave birth to the somethingness is a nothingness that matters.

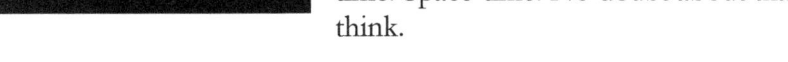

So if we blithely say that there's nothing in the universe, that nothing might actually be something.

But we need to redefine our terms. Big Bang will do that for us, since it has already shown the everything *can* come from nothing, since everything *did* in fact come from nothing.

So let's look at our questions - why is there something rather than nothing in the universe?

The first answer is, the nothing is actually something.

Here's what the universe did. We think. Maybe.

It arrived with nothing it in. No energy, no matter, no laws of physics. Just space and time. Space-time. No doubt about that. We think.

So it took no energy at all and turned it into two types of energy - positive and negative. Positive energy will quickly become matter (e=mc²), negative energy even more quickly became gravity. And they exactly cancel it each other out. The energy that produces matter, by the way, is light. Photons. Light becomes matter. All the matter of the universe comes from light.

So now we have space-time, gravity and matter.

Let's define gravity. Gravity is an interaction between space-time and matter. Matter bends space-time. Bent space-time tells matter how to move. Matter tells space-time how to bend. John Wheeler said that. He was right.

Interaction. Remember that word.

Just before the universe produced any matter, it gave us the laws of physics, the forces of nature - Strong Force, Electromagnetic Force, and Weak Force.

But as it turns out, the Forces are not with us if there are no particles for them to interact with.

So the universe gave us particles - quarks, electrons, and a sticky little particle called a gluon. It glues stuff on.

The Strong Force interacts via gluons and quarks, and glues quarks together into protons and neutrons.

Now we've got protons, neutrons and electrons. The Electromagnetic Force interacts via electrons.

Interaction. Remember that word.

That's all in about 3-4 minutes after Big Bang.

Then we wait about 378,000 years for the universe to cool down a bit. After which, it produces atomic nuclei from protons and neutrons and simple atoms from atomic nuclei and electrons. This is all because of the Strong Force via quarks and gluons and the Electromagnetic Force via electrons.

Interaction.

Gravity kicks in, and all those simple hydrogen gas molecules roll down a gravitational well or two (gravity, the interaction of space-time and matter) until things get really hot and pressured, and lo and behold, a star is born out of the interaction of space-time, matter, gravity, and a little smidge of both the Electromagnetic Force and the Weak Force (which has it own particles, the W and Z bosons).

Here's what stars do. Stars make all the elements that you need for life and planets out of hydrogen. Which is made from a proton and an electron. Which come from light. Which comes from the universe making something out of nothing. That is, matter and energy out of no energy at all. And this all happens because the universe conjured up forces.

And what is a force, anyway?

It's apparently something that makes particles do things. Or, the force is the particles, and the particles are the force.

And it all comes from nothing, except for one thing. Interaction. Remember that word.

Chapter 6   There's Nothing There – Part 4

Interaction.

The universe is a place defined by interaction between things that aren't really there.

That is, the forces of nature, and what are forces, anyway? and a handful of particles that only do anything because of the forces and according to Quantum Theory's godfather, Niels Bohr, aren't really there in the way that we think things ought to be there. Or anywhere. Which is another option - they not only aren't there, they could be anywhere or everywhere. Or not.

In fact, it seems likely that the universe began with an interaction. General Relativity tells us that the universe began from an infinitely small point called the Singularity, and Quantum Mechanics tells us that if the Singularity is small (as it was, since it had no size at all)(QM also tells us that it can't be quite that small, but, well, we don't know yet), then Big Bang had to start with an interaction. In this case, an observation of the Singularity by something capable of observing. An observer.

An observer. Since Big Bang produced Space and Time, an observer outside of Space and Time. Outside of four dimensions of Space and Time. Outside of the Laws of Physics. An observer.

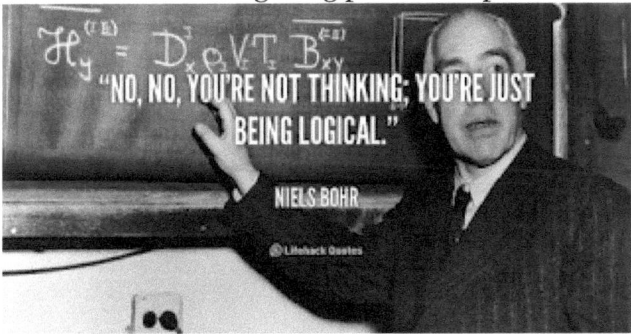

Because in Quantum Mechanics, nothing ever happens without an observation or a measurement. Nothing. Ever. Not even a universe.

I read the following the other day on Aeon.com, written by Margaret Wertheim (aeon.co/users/margaret-wertheim):

29

"Such a view appalled many physicists, who fought desperately to find a way out, and for much of the 20th century it still seemed possible to imagine that, somehow, subjectivity could be squeezed out of the frame, leaving a purely objective description of the world. Albert Einstein was in this camp, but his position hasn't panned out.

"Forty years ago, the American theoretical physicist John Wheeler proposed a series of thought experiments to test if an observer could affect whether light behaved as a particle or a wave and, in 2007, the French physicist Alain Aspect proved that they could.

"Just this April, Nature Physics reported on a set of experiments showing a similar effect using helium atoms.

"Andrew Truscott, the Australian scientist who spearheaded the helium work, noted in Physics Today that '99.999 per cent of physicists would say that the measurement... brings the observable into reality'.

"In other words, human subjectivity is drawing forth the world."

And in Scientific American magazine in May 2018: "Only in recent years has the technology necessary for answering this question become accessible, enabling a string of experimental results—including startling ones reported in 2007 and 2010, and culminating now with a remarkable test reported in May—that show that key predictions of QM are indeed correct. Taken together, these experiments indicate that the everyday world we perceive does not exist until observed, which in turn suggests a primary role for mind in nature ... According to QM, the world exists only as a cloud of simultaneous, overlapping possibilities—technically called a 'superposition'—until an observation brings one of these possibilities into focus in the form of definite objects and events. This transition is technically called a 'measurement.' One of the keys to our argument for a mental world is the contention that *only conscious observers* can perform measurements ... Recall Max Planck's position: 'I regard consciousness as fundamental. I regard matter as derivative from consciousness...' Now that the most philosophically controversial predictions of QM have—finally—been experimentally confirmed without remaining loopholes, there are no excuses left for those who want to avoid confronting the implications of QM."

Let's define what that says and means. It means that human interaction with the universe actually creates the reality that we live in. Without human interaction, there is no universe, no space and time, no laws of physics. No reality.

So the universe starts with an interaction. And then it becomes defined by interactions. Gravity is an interaction between matter and space-time. Matter is a quantum interaction between forces and particles, between the Higgs boson and energy. Stars are a gravitational interaction between matter and space-time. The elements you need for life come from that interaction, cooked up in the interior of stars or in the death of stars.

Everything in the universe, including the universe itself, is bound up and defined by interactions. You are made of stardust, made of space-time, made of particles and forces, all interactions.

Even humans and human relationships. You interact with the Strong Force and the Electromagnetic Force in order to exist. That is, your atoms do. You are made of space-time, sitting in space-time, affected by space-time, interacting with space-time. You interact with the universe - gravitationally, quantum mechanically.

And you interact with the biosphere, with plants and animals and air and water and bacteria and parasites and ecosystems and weather, as everything interacts with everything else.

And you interact with humanity. Love, hate, friendships, adversaries, random people on subways and buses and trains and planes and automobiles. With family. Loved ones.

Lovers, wives, husbands, children.

All of life is an interaction. It starts with sperm and egg interacting, preceded of course by a certain interaction between mother- and father-to-be. Embryo with mother. Newborn with mother, father, doctors, nurses, midwives. Family. Friends. All of life is an interaction, starts with an interaction, proceeds via interactions, ends with an interaction.

Some interactions are infinite. Gravity, the weakest of the universe's interactions, is infinite in its reach. The Strong Force, clearly the strongest, is limited to the size of an atomic nucleus.

At the human level, the word "interaction" is analogous to "relationship". In fact, it could be said that this is true at every level. It is an interactional universe. It is a relational universe.

At the human level, that relationship ideally starts in an act of love.

Now we make a tiny little philosophical leap. Maybe ... theological. To wit:

The highest form of relationship is love.

All of our human relationships are defined by love in its varying levels. Indifference, hatred, fondness, disgust, dislike, lust, ignorance, passion, anger; a bell curve of love. We each hate a few, we are ignorant of most, we tolerate many, we like some, we love a few. What we want most in life seems to be two things, one of which is to be loved, to belong, to be accepted, and in its purest form, this love is unconditional.

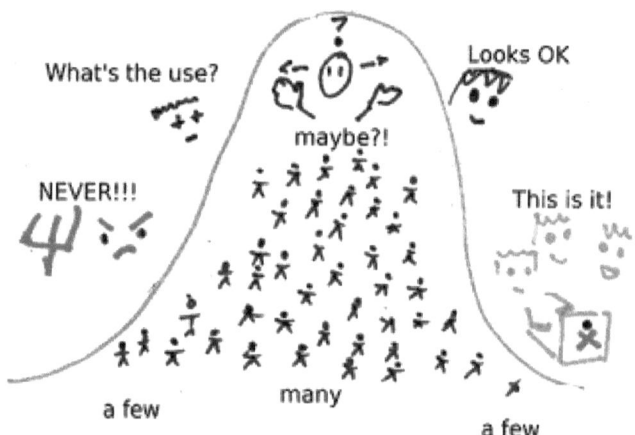

The other is that we want to matter, to have made a difference, to have left the world a better place, to have been a part of something greater than ourselves. It is probably just a variation on love - we want to be loved and admired for what we did as much as who we were.

That is, we want our interaction with the universe and our own personal version of the universe to have mattered. And we want to have loved and to have been loved along the way.

Chapter 7   There's Nothing There – Part 5

I was reading this article in Scientific American the other day all about Reality and how we don't know what it is. The conclusion reached - we can only understand anything in the universe by understanding how it is related to other things. That is, ultimately and finally, it is not just that everything interacts with everything else, but we can't understand the things themselves, which don't actually exist, mostly. We can only understand the interactions.

And the word they used in the article was ... "*relata*". Relationships.

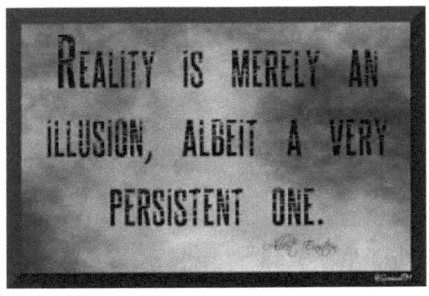

Here's what they said:

"A growing number of people think that what really matters are not things but the relations in which those things stand. Such a view breaks with traditional ... conceptions of the material world in a more radical way.

"... It runs as follows: we may never know the real natures of things but only how they are related to one another ...

"Now a question arises: What is the reason that we can know only the relations among things and not the things themselves.

"The straightforward answer is that relations are all there is."

All that there are, are relationships. It is not a universe that happens to have relationships in it. It is a universe that is defined by the relationships that are in it. It began with a relational act of interaction/observation that brought it into being, it proceeded by acts of observation that brought reality into being within it, and every action is part of a series of interactions - there are no actions in isolation; there are just interactions.

So you do not live or breathe or act in isolation. You are not alone. You are part of a series of relationships that started at the beginning of time and surround you like a cloud of angels and demons.

You might be caught up in your own existence, feeling isolated and lonely, abandoned and lost, unable to find your way, maybe in part because we have been indoctrinated to believe in fierce individualism, to believe that we must make it on our own without help or aid.

But we are caught up in relationships. They define us. They surround us. They are our curse and our salvation. They drive us to despair, they exalt us into glory. They are not just the answer.

They are all that there is.

You do not, cannot, will not exist in isolation. You exist only through and because of relationships. Richard Dawkins speaks of *"The God Delusion"*. He is wrong. The greatest delusion may be that we are individuals.

I wish then to make a proposition.

Observer God. Relational God.

The highest form of relationship is love.

There is no love greater than to give your life for others.

All you need is love, love. Love is all you need.

And maybe dark chocolate. Which is the finest sign in the universe that God exists and loves you wildly.

Chapter 8   There's Nothing There – Part 6

Here's where it gets interesting.

As if it weren't already. We might have said that once before.

There are two schools of thought, more or less.

One says, the universe is random and arbitrary and indifferent to humans, earth, the Milky Way galaxy, all the other galaxies, any other aliens that might be living out there somewhere, all the other possibly life-supporting planets that exist, any galaxies that they might be in, and itself. The universe and everything in it, including humans, is just a big accident.

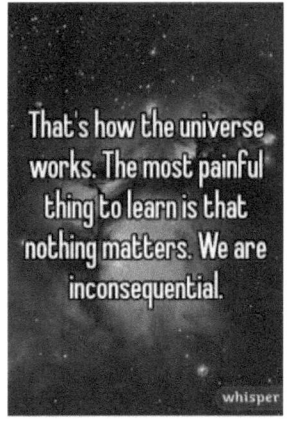

That is, nothing matters.

The other school says, God loves you and has a wonderful plan for your life.

That is, everything matters.

OK, there might be another school in the middle that says, there is no God, but everything still matters. This school is adorable, but tends to be operating out of a position that says, on the one hand, the universe is a big accident and nothing matters, but, on the other hand, I refuse to believe that I personally don't matter, dang it.

Interestingly, nearly everybody who doesn't believe in God is in the last group. I feel bad, but here's the math: if the universe and everything in it don't matter, and you're in it, then, well, there you go.

The response at that point is usually, well, dang it, I matter to me, and the universe matters to me.

Which is ... adorable. That is, terrible science, but kinda necessary in order to be able survive mentally and emotionally until, um, survival is no longer an option. A convenient, useful, not-optional delusion. To use Dawkins' word again. Turns out we need to pretend that we matter in order to stay sane and get stuff done. Otherwise, you know, what the heck is the point? So we have to act like there's a point.

Even more interestingly, actual science messes up all three groups. The way the actual universe actually is, is incompatible with any of these world-views.

OK, group the first and the second, here's your problem. Group the last, you should pay attention. The universe is not indifferent to the existence of humans or any other intelligent life form that might be out there somewhere.

The universe needs intelligent, sentient, self-aware observers in order for reality to exist.

You don't have to like that, but at the moment, it's still true.

And somehow the universe put itself together in such a way as not only to produce life, and humans, but did so in such a way that the chances of it happening by accident are mathematically and physically non-existent.

It's called fine-tuning. The universe is finely tuned.

That means that the constants of nature are each and every one of them dialed pretty much to exactly what they need to be to produce and sustain human life.

Gravity. Strong force. Weak force. Electromagnetic force. Juuuuuust right. It's so striking that Paul Davies wrote a book about it called "The Goldilocks Enigma" and another one called "Cosmic Jackpot". Because it's not only those four forces, but over 200 more constants of nature that each have to be juuuuuust right. Stuart Kaufmann wrote a book called "At Home in the Universe", and John Gribbon and Martin Rees penned another called "Cosmic Coincidences." (Neither Gribbon nor Rees are big fine-tuning fans, btw.)

Roger Penrose at Oxford calculates the odds of an ordered universe (any order, not just our type of order) appearing by accident at $10^{10^{30}}$th, and the odds of life (any type of life, not just what we've got) appearing at random at $10^{10^{123}}$rd.

That's not gonna happen.

As a way out of this conundrum, scientists have come up with the idea of a Multiverse, an infinite number of other universes outside of ours, as the only possible alternative. With an infinite number of universes, 10**10**123rd is not a problem - order, structure and life will happen somewhere automatically.

In fact (because Infinity is always bigger than you think it is), with an infinite number of universes, this very exact universe that you live in will happen not once, but an infinite number of times, and every possible variation in any kind of universe will happen not once, but an infinite number of times.

So there would be an infinite number of You out there exactly the way you are right now, plus an infinite number of You, only slightly different, each slightly different version appearing an infinite number of times in an infinite number of places.

But of course, because science is always about evidence, and there is no evidence of any other universe, the Multiverse isn't science yet, and never will be. We don't even know what most of *this* universe is like, the very universe that you live in, and never will. Detecting other universes - not gonna happen.

Apart from the fact that there will never be any evidence of even one other universe outside of ours, there is still the Observer Problem. You still need an outside observer in order for any reality in any universe to exist, regardless of how many universes you might have.

That's not to mention (although I seem to be mentioning it right now) that the idea of the Multiverse comes from three areas of physics - string theory, inflationary theory, and the Many Worlds Interpretation of Quantum Mechanics - for which there is no evidence in science. They each are pretty cool, but, well, no evidence means they aren't science yet.

Growing numbers of scientists are calling those three fields of study "pseudoscience."

Ouch. Harsh. True, though.

The only solution inside this universe is for it to be infinitely old and large.

But Big Bang messed that up nicely. That's the problem, really. Big Bang.

Before Big Bang came along, we could pretend that the universe was infinitely large and old and had always been here and didn't have a starting point and the laws of physics had always been here making stuff happen, and it was all quite lovely and pointless, just the way we like it. Didn't need any observers. Didn't even know about most of the constants that have to be juuuuuuust right.

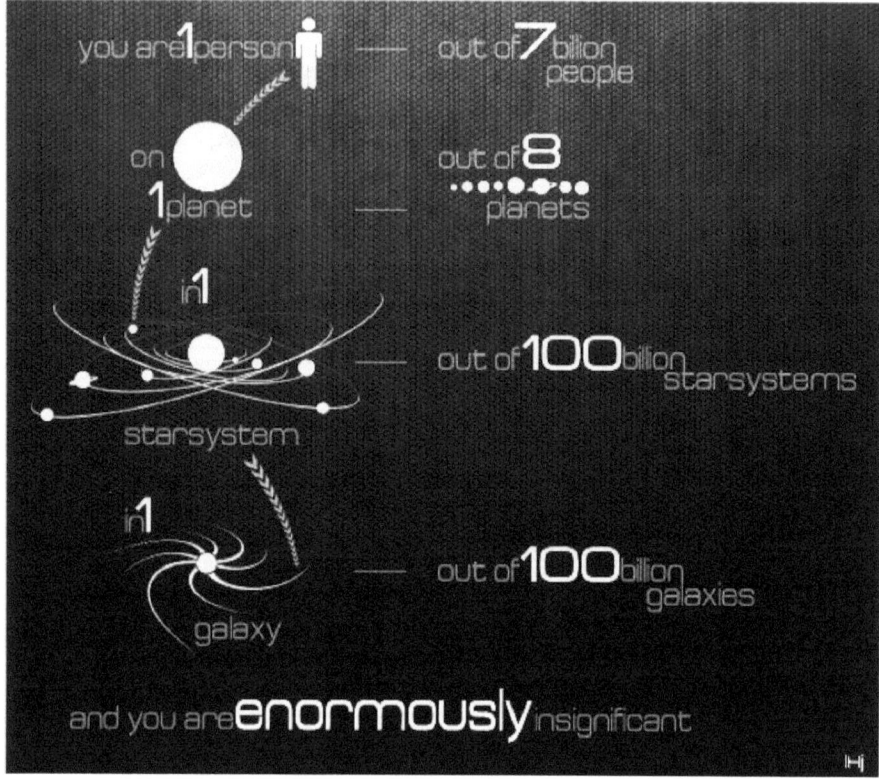

Dang Big Bang.

So your problem, groups 1 and 2, is that the universe is set up juuuuuuust right to produce humans so that they can look at it and make it be there. That's just good physics.

And what's more, it is, as we have noted, not a universe that is full of things, but a universe that is full of relationships, a universe full of things that interact, a universe where the interactions define and create reality. Reality exists in the universe because of interactions.

It's probably worth mentioning that 1) lots of smart (and some dumb) people don't like the fine-tuning argument at all and 2) make fun of people who do. Normally, people who disagree will 1) use good, real, not pseudo-science but science-with-real-evidence as a rebuttal and 2) not make fun of people. That doesn't seem to happen with fine-tuning. For what it's worth.

(The image is wrong, btw. There are probably not 100 but more like 300 billion or more stars in the Milky Way, and maybe 1 or 2 trillion other galaxies instead of just 100 billion. And many more than 7 billion people now. And that's in the observable universe. The rest of the universe may be 10^{26} times larger than the part we can see. And maybe it's infinite. Which would make you ... infinitely insignificant. Sorry. I feel bad. "Infinite" is infinitely bigger than "enormously".)

Now, group 3, you have another problem.

PS Those pointy things come from String Theory, which needs ten or eleven dimensions to work, or maybe 26, but we only see three in our universe, so the thought is that there are seven or eight other dimensions that used to be here (at the beginning of time), but aren't anymore, so where the heck are they? That's them. Little 7-dimensional lumps called Calabi-Yau Spaces, one in each elementary particle in the universe. So you've got a ton of them lurking in all of your particles, each electron, each quark, each gluon. If they exist. And we'll never know.

Chapter 9   There's Nothing There – Part 7

OK, Group the Second. Buckle up.

First, Big Bang. You need to get over it. There's an earlier chapter I wrote about Big Bang that you should go read again. Go ahead. I'll wait.

(Whistling)

(Humming)

(Foot tapping)

OK, great. Now.

The good news is that you seem to be here on purpose.

The bad news it that it's not quite the way that you think it is.

Again, let's review.

It's a relational universe. We've proposed then a relational God.

So you are bound up in relationships. That's why you're here.

This universe not only is defined by relationships, it is also *not* defined by determinism.

Isaac Newton and his contemporaries, and a lot of modern folks, thought, and still think, that it's a "cause-and-effect" universe where everything has a cause. Where nothing happens without a cause. Everything is predetermined by 1) what has come before, 2) the laws of physics, and/or 3) God, depending upon your proclivities.

So that would be a universe without 1) Quantum Theory, 2) Chaos Theory, 3) Free Will and, as it happens, 4) God.

That probably needs explaining. That might take a minute.

It is a universe that is entirely, wondrously, bafflingly, mysteriously and dramatically different than you think it is.

Here's a few paragraphs from Scientific American magazine describing the world of particles. Those are the things that you are made of, you and everything else. Get ready for some serious weirdness. If you don't get any of this, welcome to everybody else, including all (ALL!) the scientists who came up with it and fiddle with it nowadays:

"When most people, including experts, think of subatomic reality, they imagine particles that behave like little billiard balls rebounding off one another. But this notion of particles is a holdover of a worldview that dates to the ancient Greek atomists—a view that reached its pinnacle in the theories of Isaac Newton. Several overlapping lines of thought make it clear that the core units of quantum field theory do not behave like billiard balls at all.

"First, the classical concept of a particle implies something that exists in a certain location. But the 'particles' of quantum field theory do not have well-defined locations: a particle inside your body is not strictly inside your body. An observer attempting to measure its position has a small but nonzero probability of detecting it in the most remote places of the universe. This contradiction was evident in the earliest formulations of quantum mechanics but became worse when theorists merged quantum mechanics with relativity theory. Relativistic quantum particles are extremely slippery;

"... they do not reside in any specific region of the universe at all.

"Second, let us suppose you had a particle in your kitchen. Your friend, looking at your house from a passing car, might see the particle spread out over the entire universe. What is localized for you is delocalized for your friend. Not only does the location of the particle depend on your

point of view, so does the fact that the particle has a location. In this case, it does not make sense to assume localized particles as the basic entities.

"Third, even if you give up trying to pinpoint particles and simply count them, you are in trouble. Suppose you want to know the number of particles in your house. You go around the house and find three particles in the dining room, five under the bed, eight in a kitchen cabinet, and so on. Now add them up.

"To your dismay, the sum will not be the total number of particles.

"An extreme case of particles' being unpinpointable is the vacuum. Look closely at any finite region of an overall vacuum—by definition, a zero-particle state—and you may observe something very different from a vacuum. In other words, your house can be totally empty even though you find particles all over the place.

"Another striking feature of the vacuum in quantum field theory is known as the Unruh effect. An astronaut at rest may think he or she is in a vacuum, whereas an astronaut in an accelerating spaceship will feel immersed in a thermal bath of innumerable particles.

"If a vacuum filled with particles sounds absurd, that is because the classical notion of a particle is misleading us; what the theory is describing must really be something else. If the number of particles is observer-dependent, then it seems incoherent to assume that particles are basic. We can accept many features to be observer-dependent—but not the very fact of how many basic building blocks there are.

"Finally, the theory dictates that particles can lose their individuality. In the puzzling phenomenon of quantum entanglement, particles can become assimilated into a larger system and give up the properties that distinguish them from one another. The presumptive

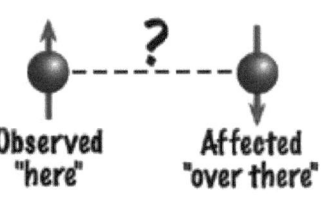

particles share not only innate features such as mass and charge but also spatial and temporal properties such as the range of positions over which they might be found. When particles are entangled, an observer has no way of telling one from the other. At that point, do you really have two objects anymore?

"... it seems you no longer have two particles anymore. The entangled system behaves as an indivisible whole, and the notion of a part, let alone a particle, loses its meaning."

So what the heck does all of that mean?

It means that the world you look at, live in, and experience is only a tiny, insignificant fraction of the way the universe actually is.

It means that when you try to fit your understanding of God into your understanding of his universe, you come up infinitely short, and make of God something that he is not.

Let's go with regular ol' Relativity for a minute. Here's the Reality of Relativity.

The faster you go, the slower time passes.

And the faster you go, the skinnier space gets.

At the speed of light, then, all of time happens at the same time. As we mentioned earlier.

And all of space is in the same place. Ditto.

So to a bit of light, the universe is a single point. It takes no time to travel from one part of the universe to the other because at the speed of light, there is no time and there is no space.

To a bit of light. A photon. But to you, since the best you can do is to be light-headed or light on your feet, the universe is huge and time passes along as it always has.

And both are true. The photon's experience of the universe is true, and so is yours. At the same time.

So here's what that means. It means that space and time, which are really just one thing together, and sort of unimaginatively we call it "space-time", are super flexible. Space-time can bend and warp, and what is really wild is this: space-time can just ... go away.

Space-time does this in two ways. One is by going really really fast, at the speed of light.

The other is when gravity gets into the act.

We should probably mention again that space-time is what the universe itself is made of. That means that the universe can just ... go away.

Hold that thought.

Chapter 10   There's Nothing There – Part 8

Wow, that's a lot of parts. Especially for a universe with nothing in it. You wouldn't think it would take 10 chapters to explain, well, nothing.

Except, as we now know, Nothing Matters. Nothing is really Something. And completely without irony, Everything came from Nothing. The second Nothing is a completely different type of Nothing than the first Nothing, to be sure. The first Nothing exists. The other Nothing, um, doesn't. Didn't. Whatever.

OK. Gravity. And the General Theory of Relativity. And the universe is wildly different than the way that anyone thinks it is.

Here's what we learned - the universe is made of space-time. Space-time can warp and bend, and for a photon, which is a little piece of light, the universe is really really different than it is for us normal, not-going-at-the-speed-of-light people. All the places in the universe are in the same place, and all the events that happen in time happen at the same time. Everything that happens, happens in the same place at the same time.

That's the Special Theory. It's not about Gravity.

The General Theory. That's about Gravity.

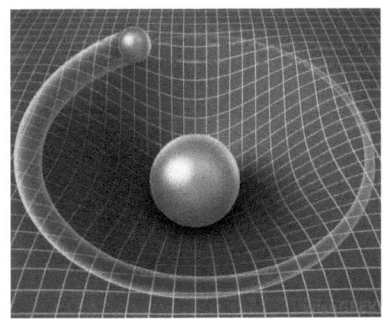

And Gravity does the same thing to space-time. It bends it. That's actually what Gravity is. Bent space-time. Looks like that over there. If you could see space-time. Kind of. Actually, it goes in towards the center of the earth from all directions. But that's hard to draw.

What Gravity is, is rolling down a hill of bent space-time. The earth bends space-time because the earth is made of matter and has mass. Mass bends space-time, creates what you might call a gravitational well. So, for example, when you jump off a diving board, you're just falling down into the gravitational well that the earth creates in space-time.

Weird. True, though.

Even weirder is this. "Down" is not "down" the way you think "down" is.

"Down" is going where time is passing more slowly.

Really.

Because the closer you get to the center of the earth (or anything made of matter), the slower time passes.

Down is not so much a space direction as it is a time direction.

Plus. Are you ready for this?

The more mass there is, the more gravity.

No big surprise there.

But. If you take a lot of mass and concentrate it into a very, very small region, space-time bends into a sphere and collapses in on itself.

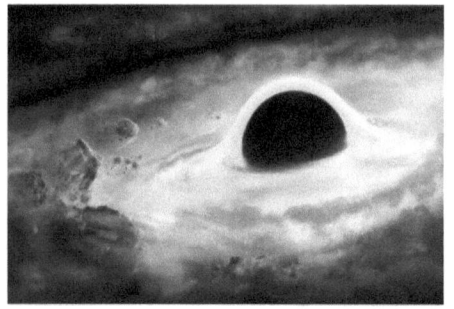

We'd call that a "Black Hole". Here's what one would look like if you could see it, which you can't, because light can't get out of or reflect off of a black hole.

And inside a black hole, space and time go away.

In fact, on the edge of black hole, which is called the "Event Horizon", it's just like traveling at the speed of light. If you went there (bad idea), you could see all of space and all of time. In an instant.

In fact again, since time stops at the Event Horizon, it would take you an infinite amount of time (we'll call that Eternity) to enter the black hole. And since Time and Space are part of the same thing, not only does Time go to infinity, so does Space.

So. The closer you get to a black hole, the farther away from it you are.

I mean, for those of us watching from outside. We would never see you go into the black hole. You would go down a time-slope into the infinite future.

For you, just like a photon, it would only take an instant, and you would see all of time and space. Maybe. We're not completely sure. You might end up painted on the surface of the Event Horizon as a hologram, along with everything else that tried to go in. And you would all arrive in the black hole (maybe) at the same time – if the infinite future can be said to be a time.

But for us anyway, you would never go in. And maybe for you, you wouldn't, either, along with everything else trying to go in.

In fact again again, what that means is that it would take an eternity for any black hole to actually become a black hole. It would take an infinite amount of time.

So there are no black holes. There are just a lot of things that are almooooooooost black holes, and they get closer all the time, but they never ever will ever become a black hole. Maybe.

For us, I mean. For the black holes, they instantly become black holes. Maybe. Off in the infinite future. Since Stephen Hawking said that black holes will eventually evaporate, they might all evaporate away before they actually got to be actual black holes. (It will take a long time – if our sun were a black hole, it would take 10^{67} years for it to evaporate. For the black hole at the center of the Milky Way to evaporate would take 10^{87} years.)

(See that Zero thing there? Einstein didn't say that. He didn't believe in Black Holes. Comedian Stephen Wright said it. Maybe.)

And this is all true. At the same time. Sort of. Since time for us is an infinity, and an infinity for black holes might be instantaneous, it's kind of hard to know how to talk about it. Plus, really, we're never ever going to really know.

So black holes are places in the universe where there are no places, and where everything happens at once, or nothing happens at all, and everything is always on the verge of happening but doesn't quite get there, or all the places eventually are the same place, off in the infinite future somewhere.

So let me say this again.

What the heck does all of that mean?

It means that the world you look at, live in, and experience is only a tiny, insignificant fraction of the way the universe actually is.

It means that when you try to fit your understanding of God into your understanding of his universe, you come up infinitely short, and make of God something that he is not.

Just in case you forgot.

Chapter 11   There's Nothing There – Part 9

So. Here's where we are.

All the scientists before the 1900s thought that the universe had always been here. Forever.

It was full of rules. And the rules made everything happen. The laws of physics and the universe - always here, always making everything happen.

Matter. Always here. Energy. Always here. Stars, planets, galaxies. Always here.

And space - just a big ol' emptiness. Time - a separate thing that they didn't quite understand, and still don't. But, you know. Ticking right along.

Everything was predictable. The laws of physics always did things in the same way, so as soon as we figured out the laws, we could predict what was going to happen.

Everything was pre-determined by the laws of physics. Everything was caused by something else. Cause and effect. That's the way it all worked.

All you had to do was to take things apart to understand how they worked. Then you would know what made them that way. And what they were going to do in the future. Galaxies. Stars. Planets. Living things. Well, dead things that used to be living, usually.

Awkward taking living things apart so that they are still, well, living after you've taken them apart.

Pre-determined. Predictable. Logical and reasonable. No weirdness. Weirdness was for religious nuts.

And nothing had a starting point. Certainly not the universe. Starting points were for religious nuts.

And everything comes from something. Always. You always gotta have something to get something else. Anything coming from nothing is for religious nuts.

And then the 20th Century arrived and messed everything up.

It's probably worth saying again that if the universe is like we all thought it was, then there would be 1) no free will and 2) no need for God. I didn't make that up. Pierre Simon Laplace did. He was a French mathematician and philosopher.

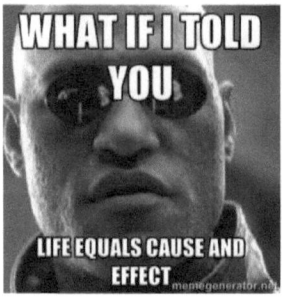

But suddenly the universe had a starting point. Space and time, all wadded up together into Space-Time, began. Then the laws of physics. Began. Space and time and the laws of physics hadn't always been here. Or anywhere.

Then energy. Arrived. And matter. Arrived.

And everything came from nothing.

Everything. Nothing.

And then everything that did arrive turned out to be nothing like we thought it was. Space and time were - Space-time. And it could bend and warp and go away and arrive in the first place. Suddenly there were places where there were no places, where all of time could happen in a single instant, where all of the places were in the same place.

And energy and matter. OMG. Particles, which were supposed to be, you know, like, just THERE doing particle things, all nice and predictable and pre-determined, were not actually THERE and sometimes like particles and sometimes like energy waves and sometimes there and sometimes everywhere all at the same time and could go from here to there without going anywhere in between and could just pop into existence

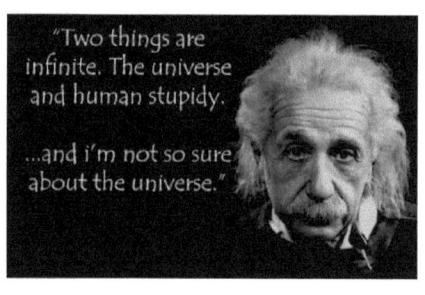

without any cause and in fact all the matter in the universe just POPPED into existence for no reason or cause whatsoever and EVERYTHING IS SO FRIGGING WEIRD. Which is for religious people.

Worst of all - nothing happens without an observation. There is no reality without an observation. And the danged universe wouldn't even be here with an observation, and is apparently so precisely wired (weird) to produce observers that it would take some sort of weird (wired) magic to set it up the way that it is set up and there's no way out except to conjure ("conjure"!?!?) up an infinite number of other universes.

Which is supposed to help. But doesn't, because we'll never know if they're there, and even if they are, so what?

And not only did everything come from nothing, but it's still mostly nothing. You could almost say that nothing came from nothing, but the now-nothing has a really strong appearance of being something. There's really nothing here, but dammit, it looks and acts like a lot of something.

But here's the reality, the coolest thing of all. It seems most likely that although there's nothing really here, that nothing has become something, not because it is something, but because of the interaction between the nothing and the laws of physics, the forces of nature. Which only exist because they can interact with the nothing.

The tiny bits of particles that aren't really there, and the vast, endless emptiness of space-time.

What's really here, then, is interaction.

Reality is reality because of interaction. Which all starts with observation. Which is an interaction.

Although it is an interaction that has to take place outside of time and space, outside of the 13.8 billion year history of our universe, outside of all of the laws of physics and outside of energy and matter.

So if we're gonna talk about God, that's where it starts.

BTW. If the universe is like this (and it is), then 1) free will and 2) God come back into the picture. Sort of, together.

Chapter 12   There's Nothing There – Part 10

There are 10 types of people in the world. Those who understand binary, and those who don't.

So. Part 10. Base 10. It's all in how the universe defines the terms.

(Isaac Newton)

So. Napoleon and this other French guy Pierre Simon Laplace were having this conversation about this book that Laplace had written about Isaac Newton's new physics. The book said that the universe operated according to laws and always had.

(Not Isaac Newton. Looks like him, tho.)

Napoleon said, so, um, there's no room or need for God in this picture, is there?

And Laplace said, nope. Don't need God. Got laws. The laws make stuff happen. Not God.

And oh, btw. If the laws of physics make EVERYTHING happen, then our brains make NOTHING happen. So there is no free will, either.

But. Of course, that assumed that the laws had always been there in an infinitely old universe.

And, as it happens, that turns out not to be true.

The universe arrived, and shortly after that, the laws of physics arrived.

So now you have two questions that you didn't have before.

One. Where did the universe come from?

Two. Where did the laws of physics come from?

And then we learned other, fascinating and disturbing things. Like, the most powerful and definitive law of physics isn't really a "law", it's a bunch of quantum mechanical equations that folks just made up that seem to work all the time and describe matter and energy and they're always right and never wrong and they tell us quite clearly that things don't always happen because some law made them happen. Sometimes things just .... happen. Uncaused.

In fact, in Quantum Mechanics, nearly everything just ... happens. Uncaused.

And Quantum Mechanics is all about matter and energy. Everything in the universe that's not nothing, in fact. Although we've already seen that everything is mostly nothing, anyway.

And so free will becomes a possibility once again.

Because if not everything is caused by a law of physics making it happen, then it's entirely possible for us to make decisions that are independent of causes and effects.

And God becomes a reasonable answer to the two questions. Not the only reasonable answer, but still, a pretty good answer.

How? you might ask Is that possible?

Well.

Perhaps you've heard of SETI? It's an organization that searches for life elsewhere in the universe. SETI stands for Search for Extra-Terrestrial Intelligence.

One way that they do this is by searching the skies for signals that might be from aliens somewhere else in the galaxy. The rest of the galaxies are too far away for us to get signals from their aliens, so we have to focus mostly on our galaxy, most of which is also too far away, but you gotta start somewhere.

How? you might ask Do they know when they get a signal from aliens?

Well.

That's a problem. Because aliens won't speak any of our languages, not even our computer languages.

So we wouldn't be able to recognize their language as, well, a language.

And. The signals will be coming as energy waves of one sort or another.

So we have to assume that the aliens are smart enough to figure all of this out, and they will try to send signals that we could recognize as signals.

Assuming we are smart, too. And that we are here, too. We would be aliens to the aliens.

So they have to send something we would recognize. So we have to look for something we would recognize.

And that would be ... go ahead, guess.

I'll give you a hint.

The laws of physics are the same everywhere in the universe.

Go ahead. Guess.

I'll give you another hint.

The constants of nature are the same everywhere in the universe.

We hope.

Go ahead. Guess.

I'll give you another hint.

A circle here is the same as a circle there.

Waiting.

OK. Time's up.

That means that the aliens could figure out what the value of Pi is. (Pi is the ratio of the circumference of the circle to the diameter.)

It's 3.14159265358979323846264338327950288419716939937510582... and it just keeps on going forever. And it's the same everywhere in the universe.

And the aliens know that, too. So all they have to do is to send it out as data, and all we have to do is recognize it.

It's pattern recognition. Intelligent life both produces and recognizes patterns. The more complicated the pattern, the more likely that it is coming from something intelligent.

That's what we assume. That's how we look for alien life forms. We look for some sort of recognizable pattern that could only be produced by intelligent life. If it's not natural constants, it's city lights or gases that don't occur naturally or the remnants of some sort of massive war (like, radioactivity from bombs) or any number of other things.

Now what in God's name you might ask does any of this have to do with God?

Well. Funny you should ask.

(PS I just have to say something about aliens here. First, it would be cool to run into them. If they didn't eat us and steal our TVs. But second, that is not going to happen. Running into them, I mean. Here's the question that reveals the why – how are they ever going to find us? Space is huuuuuge. Our analogue sound/light bubble is about 80-100 light years wide, and since we changed to digital [which is just white noise to aliens], it's not going to get any bigger. Space is huuuuuuge. Nobody will ever hear it. Nobody will ever know that we are here. There are too many "heres" in the universe for anyone to find this one. And we are never going to know if they are there, unless they are flippin' everywhere. In which case, get skinny and hide the TV.)

Chapter 13   There's Nothing There – Part 11

Pattern recognition.

That's how we measure intelligence.

The smarter you are, the better you are at recognizing patterns.

Starts early. Like, you're wandering around the jungle (maybe now, maybe 20,000 years ago), and 1) you see some tracks and 2) you hear some noises and 3) you smell something nasty, and you figure out that if you don't get the heck out of there, something large and hungry is going to 1) see your tracks and 2) hear your noises and 3) smell something tasty, and you'll end up pattern-recognized into something large-and-hungry's lunch.

All of life is recognizing patterns and learning what to do depending upon what the pattern is.

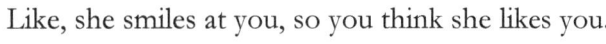

Like, she smiles at you, so you think she likes you.

Or, you have something hanging out of your nose.

So you learn to check your nose and everything else before you venture into wherever it is that someone might smile at you.

We all have different patterns we are good at recognizing.

Some are social. Some are mathematical or scientific. Some are literary or artistic. Some are political or economic.

There's a pattern - we're all different in figuring out patterns.

And the world rewards people differently at different times and places for figuring out different patterns.

The point, of course, is that life, the universe and everything are full of patterns. If there were no patterns, we couldn't figure stuff out, and we'd all get eaten by something large and hungry. Except, of course, the large and hungry would have no patterns

59

either.

So there would be nothing in the universe if there were no patterns.

Because the patterns lead us to the laws of physics. Patterns are the clues that we've used to figure out how the universe and everything in it works.

And the laws of physics are what made everything happen.

No laws. Nothing happens.

Here're more questions to ask:

1) Why did the universe do anything?

2) Why is there something rather than nothing?

And the answers are, the universe did stuff because the laws of physics arrived just after the universe did, and there's something rather than nothing (inside the universe, that is) because the laws of physics produced everything.

Energy. Matter. Particles. Elements. Heat. Pressure. Gas clouds. Stars. More complicated elements. Planets. Life. You and me.

Extraordinary, unbelievable, subtle, beautiful, and profound patterns led us to the laws of physics and to the understanding we have of everything.

Patterns - used by intelligence, produced by intelligent beings. Order, structure, complexity, beauty, elegance.

So if the universe is full of patterns (and it is), and if SETI spends all of its time assuming that if they find patterns out there somewhere that are extraordinary, unbelievable, subtle, beautiful, and profound, then they will have found overwhelming evidence of intelligent alien life, then, well.

You can do this part on your own.

OK, I'll help a little.

It means that the universe is not accidental, but is the product of an intelligence so far beyond ours as to be incomprehensible to us.

Let me quote some smart guys:

Richard Dawkins: "When we were talking about the origins of the universe and the physical constants, I provided what I thought were cogent arguments against a supernatural intelligent designer. But it does seem to me to be a worthy idea. Refutable--but nevertheless grand and big enough to be worthy of respect. If there is a God, it's going to be a whole lot bigger and a whole lot more incomprehensible than anything that any theologian of any religion has ever proposed."

Albert Einstein: The harmony of natural law "reveals an intelligence of such superiority that, compared with it, all the systematic thinking and acting of human beings is an utterly insignificant reflection."

Alan Sandage: "I find it quite improbable that such order came out of chaos. There has to be some organizing principle. God, to me, ... is the explanation of the miracle of existence, why there is something instead of nothing ... If God did not exist, science would have to invent Him to explain what it is discovering at its core."

Paul Davies: "I belong to a group of scientists who do not subscribe to a conventional religion but nevertheless deny that the universe is a purposeless accident. Through my scientific work I have come to believe more and more strongly that the physical universe is put together with an ingenuity so astonishing that I cannot accept it merely as a brute fact. There must, it seems to me, be a deeper level of explanation. Whether one wishes to call that deeper level 'God' is a matter of taste and definition." ... "To postulate an infinity of unseen and unseeable universes just to explain the one we do see seems like a case of excess baggage carried to the extreme. It is simpler to postulate one unseen God."

Antony Flew: "... some sort of intelligence or first cause must have created the universe. A super-intelligence is the only good explanation for the origin of life and the complexity of nature ... biologists' investigation of DNA 'has shown, by the almost unbelievable complexity of the arrangements which are needed to produce [life], that intelligence must have been involved,'"

Fred Hoyle: "Would you not say to yourself, 'some super-calculating intellect must have designed the properties of the carbon atom, otherwise the chance of my finding such an atom through the blind forces of nature would be utterly miniscule.' Of course, you would! A common sense interpretation of the facts suggests that a super-intellect has monkeyed with physics, as well as with chemistry and biology..."

Francis Collins: "At the most fundamental level, it's a miracle that there's a universe at all. It's a miracle that it has order, fine-tuning that allows the possibility of complexity, and laws that follow precise mathematical formulas … an open-minded observer is almost forced to conclude that there must be a 'mind' behind all of this."

Roger Penrose: "There is something absolute and God-given about mathematical truth. Not only is the universe 'out there', but mathematical truth has its own mysterious independence and timelessness."

Stephen Hawking (35+ years ago): "It would be very difficult to explain why the universe should have begun in just this way, except as the act of a God who intended to create beings like us."

Simon Conway Morris: "There is, if you like, seeded into the initiation of the universe itself the inevitability of intelligence."

There are many, of course, who disagree. I find them cranky and disagreeable.

Not stupid. They're pretty smart.

Just. Cranky.

And disagreeable.

Chapter 14   There's Nothing There – Part 12

Now.

If there is a "mind" out there, outside of space and time, outside of our 4 little dimensions, even outside of the 10 or 11 (or 26) dimensions that String Theory requires, even outside the concepts of dimensions at all ...

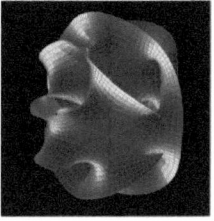

I just have to stop and think about that for a minute.

... outside of the concept of "existence" as we know it, since "existence" for us is occupying a point in space and time ...

Now I'm thinking about *that*.

... outside of the need to be created, since being created happens in space and time ...

Oh, my.

... outside of the laws of physics, since those arrived inside the universe, after Big Bang, inside of space and time ...

... not made of energy or matter, since those are part of this universe, arriving in this universe after Big Bang ...

... then the only word left comes from *The Princess Bride*.

Inconceivable.

Indeed, even much of the character of just *this* universe itself is inconceivable, much less anything outside of it. Most of it we will never

see; so much that the part we can see is almost insignificant. Space-time bends and warps and goes away and arrives. Particles are neither here nor there but everywhere and nowhere and pop into and out of existence, empty space seethes with activity, and ...

... it may be that nothing exists but interactions, relationships.

And everything came from nothing.

So we exist in a tiny little sliver of the way the universe actually is, thinking in our naive sweet arrogance that we have a clue about what's outside.

Reminds me of a guy named Antony Flew, a British philosopher at Oxford before his death not too long ago.

He became an atheist when he was 15 years old.

At age 15, he had it all figured out.

What a fine metaphor for us all.

He stopped being an atheist when he was 81. I quoted him in the last chapter.

Here's what I'm thinking.

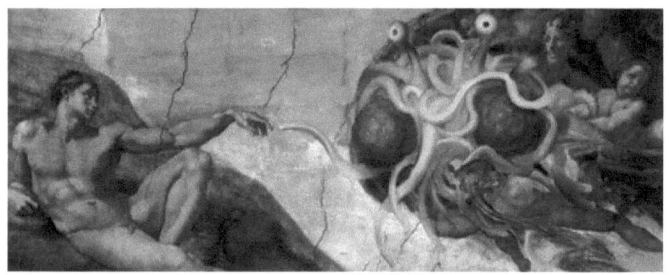

The Flying Spaghetti Monster. Look it up.

If this "mind", this intelligence, this thing that started it all isn't God, then, well, I'm a bit terrified.

If it's just an alien that's playing with us, if we are just some science experiment by a superior being, if we are just a video game being played by some super alien adolescent, then, well, I'm a bit terrified.

There are some theories floating about in science-space about the nature of the universe. Here are just three. There's lots more.

1) The universe is a hologram.

2) It's a computer simulation.

3) We actually do kind of live in The Matrix.

They're not theories, actually. They're hypotheses without evidence.

But, still. What if something, you know, just turns off the hologram generator? Or turns off the computer simulation?

We all just ... go away.

There's nothing here but interactions anyway. Those could just ... evaporate. At the touch of some switch somewhere. Dial up the Higgs. Or the Cosmological Constant. They've started calling the Higgs Interaction and the Cosmological Constant the Two Most Dangerous Numbers in Nature, because if either one of them change even a tiny little bit, we all go away. We can't do that. But the "mind" surely can.

So if it's alright with you, I'm gonna go with the Benevolent Creator hypothesis, which, since there's evidence to support it, is actually more of a theory than an hypothesis.

"Evidence?", you say incredulously. "Evidence?", you sneer. Maybe. Or maybe you're just like, evidence? Really?

Sure. Here's the short version. Everything came from nothing in a tiny tiny tiny fraction of a second.

And it all arrived incredibly precisely tuned to produce ... everything else.

And it all seems pretty well set up to produce complex life forms.

Before we knew about Big Bang, when we thought the universe had always been here, which, as it turns out, was pretty foolish considering entropy all by itself (maybe I'll explain that later), we didn't have any questions that needed answering.

We thought we had everything figured out. Lord Kelvin said in 1900 that physics was pretty well finished. We knew how everything worked. The Royal Academy of Science in London was turning down applicants, saying we didn't need any more scientists. We were done.

Then Relativity and Quantum Theory came along, with Big Bang and the obsessive need for observers.

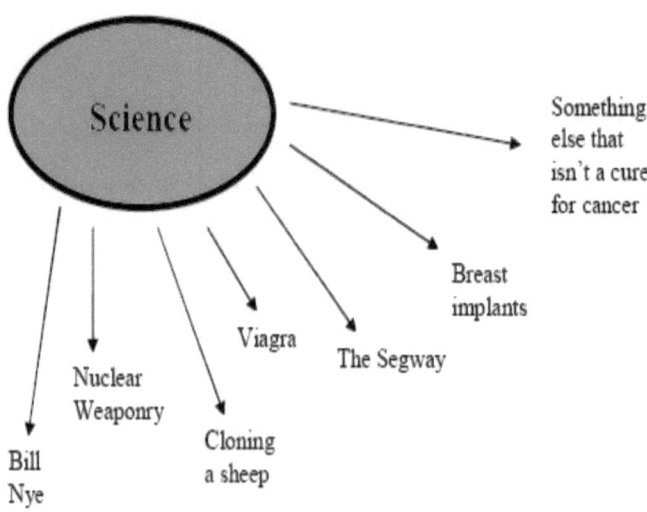

Everything pretty well figured out.

So now we have questions that we didn't have before.

And a Benevolent Creator turns out to be a reasonable answer to all of the questions. Not the only answer, but ... reasonable. Logical. Intelligent. Informed. Norant. Which I think is the opposite of ignorant. I may have made that up. Except of course you can find it on the internet.

Next, all of the questions.

Chapter 15   There's Nothing There – Part 13

So the universe just ... pops ... into existence out of nothing for no reason in the tiniest fraction of a second, and then, inside that first second, produces the laws of physics (the 4 [so far] forces [interactions] of nature), energy, particles, and inside the first 100 seconds, simple elements, and somewhere along the way, the Higgs Boson, and Dark Energy and Dark Matter. Much later, Darth Vader. And Peter Higgs. Not the same guy.

There are two interpretations of this (for us, anyway) momentous event. It was just an accident. Or it wasn't.

In the words of Douglas Adams, the universe was just one of those things that happens every now and then.

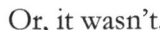

Or, it wasn't.

We'll talk about that.

First, the questions that arrived with the universe. OK, that's not exactly true. First we had to have questioners. That would be us. And then we had to discover that the universe had not always been there. Which we eventually did. And then we had to say, holy crap, now we've got some questions that need answering.

OK, then. The Questions. Some of them, anyway:

Where did the universe come from?

What caused Big Bang? Was it unique, or were there other (lots of)(an infinite number of) Big Bangs?

Where did the laws of physics come from?

What language does the universe speak? (math)

Why does the universe speak a language? (math)

Why does the universe make sense? Why can we figure things out? (math)

Why is there something instead of nothing?

Why did the universe do anything at all?

Why is there order and structure in the universe?

How does the universe go from what is presumably, right after Big Bang, a great big mess to a universe that has complex forms in it, that is, a universe that is not a great big mess? Or just empty?

Where do the elements come from?

Where did all the energy come from? And all the matter?

Where do stars and galaxies come from?

What makes all of this happen?

Where does life come from?

Is there any meaning or purpose to the universe? To life? To, well, not to sound all narcissistic and selfish, but, well, me? I mean, you know, it kinda matters. To me, anyway. And all the rest of you "me's" out there.

And the answer is ... (drum roll) ... we don't know where the universe came from (nowhere, it seems), we don't know what caused Big Bang (some sort of measurement or observation or interaction by somebody or something outside of space and time and the laws of physics and everything, but if you call it "God" it makes people nervous), and the laws of physics just ... arrived ... in the universe, and math, God (oops, sorry) only knows why math is here, and then we very much later on figured out the math and the physics (not necessarily in that order) and that gave us the answers to most of the rest of the questions. Sorta. Kinda.

Like, we don't know why the universe makes sense, but it's useful that it does so that we can understand it (via math and physics), and we know HOW there is something instead of nothing, but not really WHY except that ... (drum roll) ...

The laws of physics made everything happen.

But some things had to be reeeeeally carefully dialed in first.

For example.

entropy

The early universe had to be in the highest state of order that it would ever be in. Physicists call that extremely low entropy. "Entropy" roughly means "disorder", and the universe had to have almost no disorder in it.

Because it only goes one direction. That is, from order to disorder.

So if it had been a big mess, it would still just be a big mess and we wouldn't be here.

Like, your kitchen doesn't start out as a mess. It starts out nice and clean. And then you cook (like, a universe), and it gets messier and messier until you have two things - a lovely meal, and a big mess.

That's what the universe does. It starts out nice and clean, the same temperature (nearly), the same density (nearly) everywhere, and then as time goes on, as it gets messier, the mess produces a lovely meal. Some

parts of the universe get cooler, more dense, and other parts stay hotter (less dense), and that starts things in the kitchen right off and a meal arrives.

Well, not a meal, exactly. It produces matter first, in the form of quarks, electrons and gluons. Then protons, neutrons out of the quarks and gluons.

Then simple elements out of the protons, neutrons and electrons. Then gas clouds out of hydrogen and a bit of helium (very simple elements).

Then out of the gas clouds, stars.

And out of the stars, three things: galaxies full of stars, black holes, and more complicated elements.

And then planets, and sometime later life, and sometime after that, complex life, and then (wait for it) you and me.

We'll call all of that a lovely meal, surrounded by a universe that is much messier than it use to be. Much more "entropic". Disordered. Thermodynamically speaking.

But somehow, because the universe started with very little disorder (entropy) in it, it produced a lovely meal.

And you start to wonder ... how exactly did that happen?

Because (let's be honest), lovely meals don't become either lovely or meals without a cook, maybe some recipes, surely some ingredients, some heat, a bit of stirring, and time. And a kitchen. We'll call the kitchen "space-time". So?

And the answer is ... the laws of physics made it happen. Those would be the recipes, sorta.

But only, as we've been told, if the recipes are juuuuuuust right.

Goldilocks. Coming up.

Chapter 16   There's Nothing There – Part 14

So. Here's what you need to make a universe with something in it.

First, a universe. Any size. Not very old. In fact, brand new. Right off the lot. Very very low mileage. Made of Space-Time. Really really big, though. Huuuuge. Enormous. So when we said "any size" right back there, it was kind of a Henry Ford-ish "any size". Ford said you could have any color Model A car that you wanted, as long as it was black. You can have any size universe you want, as long as it's huuuuuuge. Just like everything else, though, it has to be exactly the right size of huge.

With nothing in it. That's what we got. An enormous universe with nothing in it.

So we need to find a way for a universe with nothing in it to become a universe with something in it.

By the way, when we say "something", we mean, like, order. Structure. Organized things.

Since we only have our universe to look at, we're going to have to go with what happened in our universe. Odds are good anyway that 1) there aren't any other universes and 2) if there are, most of them will have nothing in them. That's what we're told. Unless there's an infinite number of them. Then, 3) most of them will have nothing in them. 4) In an infinite sort of way.

Anyway. Second, rules. Like, the laws of physics. So our universe produced those. Gravity was first. Then the Strong Force. (Gravity is the weakest. The Strong Force is [duh] the strongest.)

Then electromagnetism and the weak force, which arrived simultaneously because right before that, they were called the electroweak because they were together (That would be an awesome name for a superhero - "Electro-Weak!").

71

Right before that was called the Grand Unified Theory because the Strong Force was together with them. We'll call it the Grand Unified Theory "GUT", because, um, that's what they call it.

Right before that was called the Theory of Everything, because all four forces were together. Except Gravity isn't a force. Don't worry about that. It's just confusing. But we'll call the Theory of Everything "TOE", because, well, that's what they call it.

It's too bad we don't call them Big, cuz then we'd have the Big GUT and the Big TOE. And that would be funny. Ha.

Worth noting: They are all interactions. Gravity is an interaction between matter and space-time. The others are often called the Strong, the Weak and the Electromagnetic Interactions.

Here's a good question. Why or how did the universe just ... come up with ... laws? I mean, what's up with that? Weird enough that you get a universe at all, but then the laws of physics that make everything happen just ... appear? All of a sudden, we got laws? Very strange.

Anyway. Third, energy.

OK, now we have a problem. Where the heck do you come up with energy when there isn't any?

Hmmmm. We're going to have to get creative.

Here's what we'll do. We'll take no energy, which is what we have, and divide it into two.

Hmmmm. How exactly does one do that?

Well. Since we're good at math, and the universe is all about math, we'll divide no energy into a positive half and a negative half. And they'll just exactly balance each other out. Which makes zero.

So once again, we take ... nothing ... and turn it into ... something.

The positive half we'll call ... matter. And the negative half we'll call ... gravity.

And they'll just exactly balance each other out.

Genius.

Wait. How are we going to turn energy into matter? I mean, you can't just ... *do that* ... can you?

Well. As it turns out, we can. It's called $E=MC^2$.

But you might recognize it. Albert (The Man) Einstein came up with that. And what it means is ...

Energy just becomes matter.

In fact, since the whole electro-magnetic spectrum is just different forms of light, what really happens is ...

Light becomes matter.

And so it did.

Why? How!? What's up with that!?! Light just ... BECOMES MATTER?!?!?

Well. Yes. We have a law for that. $E = MC^2$. There you go.

Let's review.

1) The universe comes out of nothing.

2) The laws of physics come out of nothing.

3) Energy comes out of nothing.

4) Matter comes from energy, which came out of nothing.

5) All the matter is made of light.

6) And the whole thing starts with almost zero entropy. That is, in almost the most perfect state of order that it would ever be in.

So, thus far, what we've got it is, nothing. That is rapidly becoming something. Sort of.

Chapter 17   There's Nothing There – Part 15

Now, the problem with a universe, a brand new universe full of Space-Time, or made of Space-Time, is that it is a bit unstable.

Like, brand new babies. Like, here's your brand new baby. Isn't it cute? Don't you just want to cuddle it to pieces?

And then you find out that it poops and pees and cries and screams and squirms and doesn't sleep at night when YOU need to sleep, but it sleeps really well in the daytime when YOU need to go do stuff and grows up and becomes a screaming toddler and then suddenly a nasty old teen-ager who is completely and totally unstable and says that it hates you even though it really doesn't mean it. But, still. It hurts.

OK, that's not really a great analogy, but we got to use the word "unstable" in a sentence, because that's what a brand new universe might be.

Instability in a universe - very bad.

Because the universe could just ... collapse back in on itself and become, um, not a universe again. Nothing there. Not a nothing that might be mistaken for a something, but like the original nothing out of which the universe came in the first place.

That would be bad. No universe, no nothing, no order, no life, no you or me. No screaming babies or sullen teen-agers. Don't go there. Don't wish the universe would go away just because of a nasty old teenager. Tempting though it may be.

Or the universe could expand waaaaaay too much, and then you get an even more enormous, but entirely empty universe.

Because everything that the universe produces has to be close enough to everything else the universe produces in order for the universe to produce anything else, up to and including babies and teen-agers.

Because it seems likely that what there is in the nothing that makes the nothing act like something is (wait for it) ... interactions. Relationships.

And for things to interact, they have to be close enough together for the laws of physics to make them interact.

The laws of physics being Gravity (which is really weak and if things aren't close enough, Gravity is just useless, even though its reach is infinite.), and Electromagnetism (ditto), and the Strong Force (OMG, it only reaches across an atom, which is veeery small) and the Weak Force (which we personally still don't quite get yet, but take our word for it - things gotta be close).

Just to remind us - Gravity is an interaction between matter and space-time.

So here's another question: how hard is it to get a universe to be juuuuuust right? Like, it doesn't collapse and it doesn't get too big. Not too small. Not too big. Juuuuuust right. Goldilocks-ishly. We made that word up.

And the answer is (drum roll) really really hard.

If we were to give you the odds, it would 1 in 10^{60}, which is 1 in 10 to the 60th, which is 1 in

1000000000000000000000000000000000000000000000000000000000000th.

If we counted all the zeroes right.

Gravity has to be exacccccctly right in order for that to happen. Well, it can be off by 1 part in 10^{60}.

And it also has to be what we call "Flat", this universe. Which means it can't be bent one way or 'tother.

Those are the three possible shapes of the universe. Right there you can see them with The Physics Girl.

The scientists call this "The Flatness Problem". You could look it up. Good idea. Go do that.

Because it's a problem. The odds are not good. The

universe should have been bent one way or t'other.

But it isn't.

What are the odds?

That would be 1 in 10^{62}, which is 1 in 10 to the 62nd, which is 1 in 100000000000000000000000000000000000000000000000000000000000000th.

If we counted all the zeroes right.

So the scientists looked around for a way around this problem. Three of them found a way. If you want to know their names, you can ask us and we'll tell you.

They came up with this idea called "Cosmic Inflation". It's very cool. It's still pseudoscience, but it's very cool pseudoscience.

It says that the universe, which was already expanding pretty darn fast, all of a sudden started expanding super-dooper fast for a really really tiny amount of time.

We told you this before, like, twice, but that's OK. We'll tell you again.

It expanded in 10^{-35} of a second from the size of a nanometer to 250 million light years across. And then it all of a sudden slowed down again.

Here's how fast: .000000000000000000000000000000000001 second.

Here's how big: call it 2,500,000,000,000,000,000,000 km.

The Milky Way, by the Way, is only 100,000 light years across. Like, 1,000,000,000,000,000,000 km. Way smaller. Waaaaay.

(Sometimes we use miles and sometimes kilometers. Just because.)

Cosmic Inflation. It's a lovely idea.

With a problem.

Things needed once again to be juuuuuust right for Inflation to happen.

Goldilocks-ishly.

So in order for the universe to be just right in order for it to actually be able to do anything, it had to expand super-dooper fast for a tiiiny amount of time, and in order for that to happen, everything had to be juuuuuust right.

And then there are some other things that had to be pretty darn perfect. The Higgs, the Dark, and the Dark. Boson, Energy, and Matter, that is.

We should explain. We'll do that. Later.

Chapter 18   There's Nothing There – Part 16

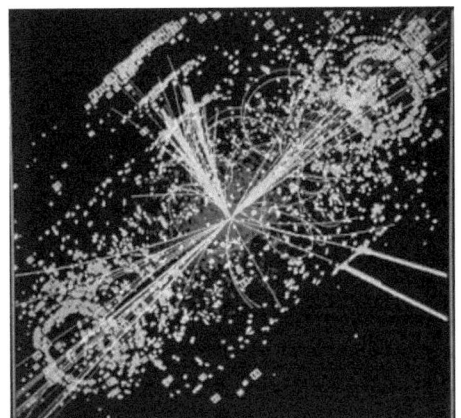

The first thing to say about the Higgs Boson is that we also call it the Higgs Interaction. And the Higgs Field. But we're gonna go with Interaction, because it fits so nicely with the whole Interaction thing. It's in that picture somewhere. So I'm told.

The second thing to say is that when I tried to give a talk which included a tiny little bit of talking about the Higgs to a group of people that included some of the scientists who found it (this was in Geneva, very close to CERN, the particle accelerator where the Higgs was found), I didn't do so good.

I don't know why. I looked up the Higgs on Wikipedia the night before just to make sure. This is true.

So if you go look up the Higgs on Wikipedia (which you should do right now), here's what you'll find out.

It's way more complicated than anything I'm gonna say here.

So I'm just gonna go with what I read in the papers.

It's not that I'll be wrong. Just way too simple.

Like describing a rainbow as "Pretty. With colors."

So if you're a physicist who helped discover the Higgs, you could just stop reading here and skip ahead or go drop apples out of trees or something. Go reconcile Relativity with Quantum Theory. That would be nice. Or figure out Dark Energy and Dark Matter. You could win a swell prize.

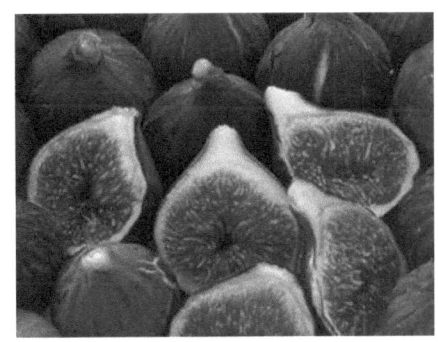

Not the Higgs

Anyway.

Nope. Still not the Higgs

Right after Big Bang, we had all these particles whizzing about and all these laws sitting there ready to go to work, but here's the thing.

That work is "interaction".

And in order for the interaction to happen, particles generally have to have "mass".

Mass is what bends space-time, and that's what we call gravity. No mass, no gravity. No gravity, no stars or galaxies or elements or planets or life. Or you and me. Or Mass. If you're Catholic. So, really, no mass = no Mass. Little physics/Catholic joke. Sorry.

Plus, gravity needs particles that interact; gravitons. We haven't found them yet. That's because they're really hard to find. And might not exist.

Also, the Strong Force needs particles that interact. They're called Gluons. Because they glue stuff on. Gluons. They glue quarks together to make protons and neutrons, and then protons and neutrons together to make atomic nuclei. No Strong Interaction, no atoms, and no life. Gluons are bosons. No mass. Quarks have mass. Fermions. Gluons interact with quarks. All bosons are glue.

Gravitons and gluons are massless, btw. They still interact.

And the Electromagnetic Force (also called the Electromagnetic Interaction) needs particles that have mass.

Electrons. Duh. No Electromagnetic Force, no atoms, and no life. It also needs particles that don't have mass, but do interact - Photons. Light particles. Bosons.

No. Really. This is jigs. Not Higgs.

And the Weak Force needs particles. We call those the W and the Z bosons. The Weak Force is also called, as above, the Weak Interaction. The Weak Interaction, along with Gravity, makes stars, and inside stars, the Weak Interaction makes elements. No Weak Force, no elements, and no life. W and Z are (obviously) bosons. No mass. They interact with things that have mass.

So some guys (Peter Higgs was one of them. There were five others. Go look it up.) figured out that in order for particles to have mass, there had to be another particle that would create an Interaction that would give mass to all the particles that need mass. This was, like, 60+ years ago.

Wigs. Wigs. Not Higgs. Sheesh.

It took a long time to find it. And a lot of money. A crapload of money. I mean, a craaaaplooooad. Over $13 billion. To find a particle that exists for a ten-sextillionth (10^{-22}) of a second. A thousandth of a billionth of a billionth of a second. And even then, you don't see the Higgs. You see all the particles it decays into. So you get evidence of the Higgs, not the Higgs itself.

And the universe is chock full of Higgs Bosons. Everywhere. All the time. Wave your hand and knock them around by the quadrillions. But we couldn't find it without spending over $13 billion and spending 60 years at it. And it took, like, 3000 scientists. Maybe more.

It's a funny place, the universe. But it's gotta have mass in it for anything to happen, and so, the Higgs.

It does a couple of other useful things.

One. It keeps the universe from decaying. Which, as it turns out, the universe could do at any moment if the Higgs decides to change its value by a tiny little amount. Then we all just ... go away. And the Higgs could actually do this. I wouldn't worry about it. You'll never know. You'd just ... go away.

Two. It solves a problem we have with anti-matter. And matter.

Short version is, the universe is always creating matter and anti-matter particles which immediately destroy each other. Seems a bit pointless.

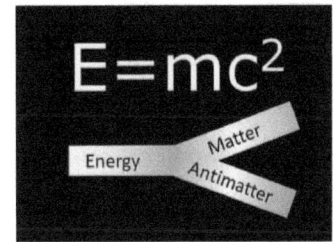

And it would be, except that's where the particles all come from.

No particles means no anything. Gotta have particles.

The Higgs arranges it so that every now and then (once in 10 billion times), a particle arrives without its anti-particle. So, we get particles. With mass. In a universe that doesn't collapse and go away.

(Einstein thought it was only 1 billion. Oops.)

Dang. That Higgs. It's really something. If this is true. We're not sure yet. At all.

Here's something else. "The tiny little mass of the Higgs boson, whose relative smallness allows big structures such as galaxies and humans to form, falls roughly 100 quadrillion times short of expectations. (10^{17})." Change it just a little, and the universe has nothing in it. That's from Quanta Mag.

This is from ScienceAlert:

"According to Einstein's theory of general relativity and the theory of quantum mechanics, the Higgs field should be performing one of two tasks.

"Either it should be turned off, meaning it would have a strength value of zero and wouldn't be working to give particles mass, or it should be turned on, and, as the theory goes, this 'on value' is absolutely enormous, But neither of those two are what physicists observe.

"In reality, the Higgs field is just slightly on. It's not zero, but it's ten thousand trillion times weaker than its fully-on value - a bit like a light switch that got stuck just before the 'off' position. And this value is crucial. If it were a tiny bit different, then there would be no physical structure in the Universe.

"Why the strength of the Higgs field is so ridiculously weak defies understanding."

Here's something just from me. I'm gonna make it up. Let me know if it's wrong.

Photons of light have no mass. So they travel at (duh) the speed of light. The speed of themselves.

So (here's my thought) if all the other particles have no mass, then they could travel the speed of light. I don't know if they would. But they could.

But for photons, the universe doesn't exist like it does for you and me and things with mass. For photons, everything happens in the same time and at the same place.

So if particles had no mass, wouldn't everything for everything happen at the same time and at the same place?

Which would mean that the universe would be just a single spot of Space-Time. Because the universe is a Higgs Interaction between particles and Space-Time.

If there was no mass. No Higgs. No mass. No universe. Just something very like the Singularity. Not exactly like it. Just ... close.

Just a thought.

Mr. Biggs. NOT Mr. Higgs. Still. The Higgs is Always Good.

Chapter 19   There's Nothing There – Part 17

Alrighty then. If the Higgs turned out to be what we thought it should be, then the universe should have ripped itself apart. We were off by 10^{17}, which is 100 quadrillion times smaller than we thought it should be. Not only is the Higgs exactly what it needs to be for us to be here, even the universe wouldn't be here if it were different.

That's a big mistake. Biiiig. Huuuuge. Like, you got your first paycheck, and instead of it being $1 (which would be deeply disappointing for a paycheck), it was $100,000,000,000,000,000. Only the other way around. Which would not be disappointing at all.

I googled how much money is in the whole world right now, just for comparison. The largest amount I got was $241 trillion. That's about 500 times smaller than your first paycheck. Good luck at the ATM with that. You'd need 1000 of the banknotes below. If they were worth

anything.

But that's not the biggest mistake we ever made in physics.

The biggest mistake was what we thought the Cosmological Constant ought to be.

What, you just said out loud, the heck is the Cosmological Constant?

I'm sure you said "heck". This is a family science book. That occasionally says things like "crapload".

Anyway.

I'm so glad you asked what the Cosmological Constant is. There's a good story.

So. Albert Einstein came up with the General Theory of Relativity. It was awesome. And it was always right.

85

Then this Belgian guy who was not only a brilliant physicist but a priest (seriously!?) was working through the math one day and discovered something. His name was Georges Lemaître. You should pronounce that in the French way. I'll help. It's pronounced "baguette". You're welcome.

He discovered that the math of the General Theory predicted, all by itself, with no evidence at all to support it, that the universe was not infinitely large or old but actually had a starting point.

And the General Theory was always right.

This was upsetting.

Everyone was busily assuming that the universe was infinite in size and age, that it had always been here.

But it hadn't.

Albert didn't like that at all. So he figured that he needed to fix it, because everybody (I mean, everybody) knew that the universe didn't have a starting point. Except religious people. And you know how *they* are.

So he tried to fix it. He stuck this number in his beautiful, gorgeous equation to try to force the universe to be infinite. And the number was called (drum roll) ...

The Cosmological Constant. He gave it a Greek letter, lambda (Λ), and just kinda stuck it in there.

$$R_{\mu\nu} - \frac{1}{2} R g_{\mu\nu} + \Lambda g_{\mu\nu} = \frac{8\pi G}{c^4} T_{\mu\nu}$$

It was ugly. Here's what it should look like instead:

Much better.

He left it there for awhile. Hoping.

And then Edwin Hubble (who invented hubble gum)(I may have made that up)(there's no such thing as hubble gum) discovered that the

universe actually was expanding, and therefore was not infinite. That it had a starting point. He found the physical evidence that this was true.

Space-Time (you remember Space-Time?) was stretching as time went forwards, so if we look backwards in time, Space-Time is doing the opposite of stretching.

Unstretching. Contracting. Getting smaller. So small that eventually all of Space-Time just ... goes away. So then, if you start right there at the beginning, the universe just ... arrives.

Later on, Fred Hoyle called this "the Big Bang", 'cause he thought it was a dumb idea.

But it wasn't. And all of the religious people were generally right about the whole beginning of the universe thing.

So Albert had to take the Cosmological Constant out of the General Theory. It was a bit embarrassing. He had to apologize to Lemaître. In French, no less. For a German, that really hurts.

That was, like, 1929.

Time passes. Einstein dies. Hubble dies. Lemaître dies. Not necessarily in that order.

Nobody pays very much attention to the Cosmological Constant. It was a bad idea. Very bad. Totally wrong.

Hah! is what the universe says about that.

Then in 1998 (almost 70 years later), some guys discovered that the universe was expanding a lot faster than we thought.

Huh, they said. What would make that happen?, they wondered. So they said to themselves, in order for the universe to start expanding faster than it was, that would take something that looks a lot like ... energy. Like, the opposite of gravity. Gravity draws things together. This is things getting pushed apart. And that would take some sort of energy.

Huh, they said. Why can't we see it? What the heck is it? Where the heck is it?

So they won the Nobel Prize. But they still don't know what it is.

So they called it ... Dark Energy. Which means, we don't know what it is or where it is or how it works, but it kinda seems like it's got to be energy, but we can't see it or measure it or touch or anything.

So we'll call it Dark Energy. "Invisible, Undetectable, Immeasurable Energy" would have been better, but Dark is like, a cool word. They say that Dark Energy might be the energy that empty space has.

Huh?

So empty space now has invisible energy in it?

Very strange.

And guess what they decided to use? Einstein's Cosmological Constant.

And they figured out what the value would have to be. And they wrote it down.

And they were so, so, so wrong.

The number they predicted is a trillion trillion trillion trillion trillion trillion trillion trillion trillion trillion times bigger than the actual value. That's 10^{120} bigger.

And if the Cosmological Constant was any bigger than it is, the universe would be ripped apart. Again.

That's the biggest mistake in the history of mistakes. Here's how big that is:

There are about 10^{120} elementary particles in the entire observable universe. Plus or minus. We were wrong by the number of particles in the universe. That's a crapload of wrong, is what that is. It's a thousand trillion trillion trillion times bigger than the number of atoms in the Universe.

So here's what we've got. If either the Higgs Boson or the Cosmological Constant were what we thought they should be, the universe would have blown apart before it ever got started.

They have to be just what they are. Or we wouldn't be here.

They've been called "The Two Most Dangerous Numbers in the Universe."

Here's a quote:

"At the core of Cliff's argument are what he calls the two most dangerous numbers in the Universe. These numbers are responsible for all the matter, structure, and life that we witness across the cosmos. And if these two numbers were even slightly different, says Cliff, the Universe would be an empty, lifeless place." Harry Cliff. He's at CERN.

And it's the Higgs *Interaction*, of course. And the expansion rate of the universe is an *interaction* between Dark Energy and Space-Time. Because all there is, is *interactions*.

Gotta have the Higgs. Gotta have Dark Energy.

Next. Gotta have Dark Matter.

Chapter 20   There's Nothing There – Part 18

Fritz Zwicky. Making science. OK, not Fritz.
Some Swiss guy on an alpenhorn. Making noises

There was this guy named Fritz. Fritz Zwicky. He was Swiss.

My sources tell me that he was ... irritating. Obnoxious. Unpopular. Curmudgeonly. That's the word Wikipedia uses.

Actually, he was lucky. If he'd lived in an English-language based country, he'd have been called Icky Zwicky. Or worse. That would have made anybody irritating. Actually again, he lived a long time at CalTech in Pasadena, CA, where the natives speak English after a fashion, so "Icky Zwicky" might have been spray-painted on his locker, maybe.

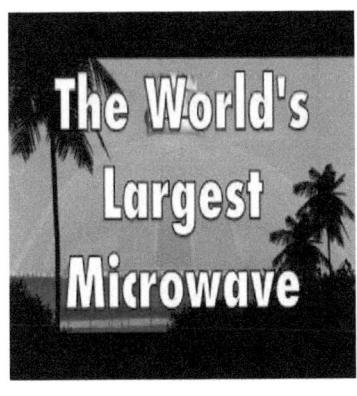

Not really. That's at CERN.
This is ... not at CERN.

I lived in Switzerland. Three times. When people ask me sometimes how come I lived in Switzerland three times, I say that it's because God loves me more than he loves them. Cuz Switzerland is very cool. Alps. Cheese. Chocolate. Wine. Skiing. Biking. Hiking. Not a big hiker, but if I were, I'd hike there. Plus, you know. CERN. Particle physics. The Higgs. The World's Largest Microwave

91

Oven. Actually, CERN is mostly in France, but there's a little piece in Switzerland, so we're gonna go with it.

Anyway. I digress. Fritz. You could rhyme something with that, too, that we won't do because "crapload" is kind of our swearing limit.

Fritz. Way back in the '30s, right after the General Theory kicked in, Fritz was looking at galaxies (which we didn't even really know for sure existed until like 1927 or -8 or -9), and, being a smart guy, noticed that they were spinning too fast to be able to hold themselves together.

You know how if you get on a merry-go-round (do we still call them that?), or a thing that you go around and around on so that you can get merry, but you up your chuck instead? Or you fall off, if you're not holding on? Or both?

Well. Galaxies are spinning so fast (not that you can tell by just looking) that they would throw all of their stars out into space, and they wouldn't technically be galaxies anymore, since, um, you gotta have stars to be a galaxy.

So he (Fritz) wondered, what was holding the galaxies all together?

Well, the answer is, gravity. Gravity is what holds galaxies together.

That is, all of the gravity in all of the stars in each galaxy is added up and holds the galaxy together. The stars make a gravitational well, and the galaxy sits in the middle of the well.

But he (Fritz) noticed that there weren't enough stars in ANY galaxy to do that. At least, the ones that he could see. So there wasn't enough gravity in ANY galaxy if you just counted up all the stars.

We needed more gravity.

And what makes gravity, you ask?

Stars. That is, matter. Stuff made of matter.

So we needed more matter in every galaxy. EVERY galaxy.

How much? you might ask.

About 5-6 times more matter than is actually there.

5-6 times more stars than are actually there.

That would be like, in our solar system, we'd need 5 or 6 more suns, plus maybe instead of 8 planets, 40 or 48 more planets.

Now, if you'll go out and look around, you'll notice that we have one sun, tops. And rumor has it that we only have 8 planets. Sorry, Pluto. Maybe 9 if the mystery planet turns out to be a planet.

So there is a crapload of matter missing in the universe.

We could have known this in 1933 when Fritz discovered it, but he was kind of irritating. So we didn't pay attention.

Darth Matter. Not Dark Matter. Get it? Get it?

And then, in 1975, a much nicer scientist person named Vera Rubin discovered it all over again. She was just as smart, but not such a pain. Not a pain at all, in fact.

She should have won a prize. Not for being not a pain, but for discovering all over again that a crapload of matter was missing from the universe. And then she passed away before they could give her the prize. Dang. That's sad.

Fritz decided to call it "Dark Matter". Since it's missing, we have no idea what it is, don't know what it's made of, and can't quite figure out how SO MUCH MATTER is out there that we CAN'T FIND.

We're not upset. OK, maybe a little.

Worse yet, it turns out that we really really need Dark Matter in order for the universe to work.

Because without Dark Matter, there would be no galaxies, and without galaxies, there would be only isolated stars, and for various reasons, no planets and no life. Anywhere.

No Dark Matter = no galaxies = no life = no you, no me, no us.

No Dark Energy = no universe = no life = no you, no me, no us.

So, you might ask. How much Dark Energy and Dark Matter do we need? In order for everything to, you know, be here?

Ah. Good question.

Here's what we can see. Matter, which makes up, um, everything. Stars. Planets. Dust. Galaxies. Everything we can see.

That turns out to be ... 5% of the universe. + or -.

The rest of it - 95% or so - is Dark. Energy and Matter. 68% and 27%. In case you were wondering.

95% of the universe is invisible to us. A mystery.

Because it only interacts gravitationally. So we can't see it. Touch it. Hear it. Taste it. Smell it.

It's all around us, but it doesn't interact with us except via the very very weakest of all the forces, gravity.

Dark Matter. Dark Energy. Gotta have 'em in order for us to be here. Don't have a clue what they are.

Oh. Here's what else.

This might be true. Or not. Hard to tell.

But. Since Dark Matter is, after all, Matter, and since the non-Dark Matter that we can see started out as particles, and then became parts of atoms, and then became atoms, and then became dust clouds, and then became stars, and then became 1) elements and 2) galaxies, and then became 1) planets and 2) life, and then became 1) you and 2) me, then ...

Dark Matter might have done the same thing.

So. There might be Dark Galaxies, and Dark Planets, and Dark Life, and ...

Dark You. And Dark Me.

There might be a whole Dark Universe with Dark People in it, or Dark Aliens, that we can't see because it's, you know, Dark, and ...

It's 5-6 times bigger than the universe we do see.

And to them, we're Dark, too. And they're all around us, and we're all mixed up inside of them. And they can't see us and we can't see them. But we know that there's something there that we can't see. And maybe they do, too.

There might even be Dark Black Holes. Which are even more invisible.

What a great universe!

Not Dark Aliens. Darkish.
Not a merry-go-round.
Dark aliens tend to get nauseated
on merry-go-rounds.

Chapter 21   There's Nothing There – Part 19

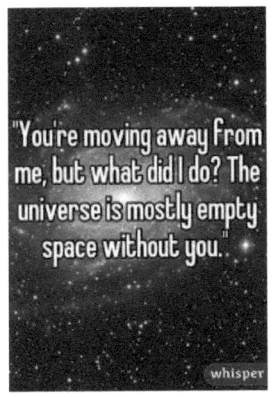

We should probably sum things up a bit, since I don't even remember where we were and I'm not really sure where we are, and I've been writing this thing for months.

We started off with, There's Nothing There, because there's nothing there. That is, in the universe.

There's a lot of stuff that looks like something - you know, galaxies, stars, planets, used-to-be-planets, might-be-planets, Black Holes that aren't quite Black Holes yet, moons, asteroids, comets, meteors, meteorites (which sounds very much like little meteors to me), plus all the stuff on earth like, well, you and me - but as it turns out, all the stuff is made of matter, and matter is made of things that - aren't really there.

That's because matter is made of energy, which is waves, and sometimes matter acts like matter, and sometimes it acts like waves.

And it does this depending upon how we choose to look at it.

And the little matter bits could be here, or they could be anywhere.

And even if the little matter bits were here, there's a lot more of nothing than there is of them.

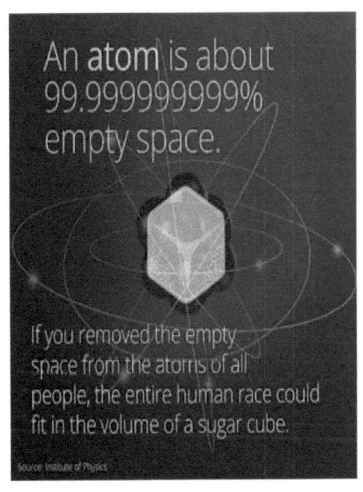

Like, there's a million billion times more emptiness in every atom than there is little matter bits. And you are made of atoms. Which means that there's a million billion times more nothing than there is something in you.

It says sugar cube. It's really an M&M.
In Europe, a Smarty.
In Asia,
a grain of rice the size of an M&M.
Or a Smarty.

You might remember we pointed that out already.

97

And 99% of THAT is virtual particles that are only there very very briefly.

And the 1% of the M&M that is actually made of little matter bits - is made of things that aren't really there.

So the way that we fix this is by saying that instead of the universe being made of matter, we say that the universe is actually made of ...

Interactions.

And what, you are saying out loud, the heck are "interactions"?

Well. Interactions are what forces and particles do together when they do what they do together.

I'm thinking that didn't help much.

OK. So, say you throw a baseball. For those of you who live in countries without baseball, pick the ball of your own culturally appropriate choosing. Cricket -. Foot -. Basket -. Human head -.

You need to thank God that we discovered baseballs and footballs and all that. Bad enough to be the geeky little kid who always got chosen last. Much worse to be the one kid left over when someone yells, hey, we don't have a head to play with! Somebody get me a head!

Not a good look when they're looking for a head.

Anyway. You throw a ball. At the simplest level, there are two things that need to interact in order for you to throw a ball.

Thing 1 - you. Thing 2 - a ball. Without both you and the ball, the ball doesn't get thrown.

We need some other things too, of course. Gravity. Which means, a planet. Various laws of physics - momentum, acceleration, deceleration, force, mass, velocity, all kinds of things.

So the whole throwing-of-the-ball-or-head thing works because of interactions between you, the ball, the planet, air, wind, altitude, mass, all kinds of things. Plus, your own desire to throw the ball, your knowledge of how to throw the ball, and your body's instinct or training on how to throw the ball.

And your body and the ball being made of particles and forces that interact.

Particles = quarks, gluons and electrons.

Forces = strong force and electromagnetic force.

Everything interacting with everything else.

That's how the universe works.

Except for this one, new thing. It turns everything around upside down and backwards.

It's not you and the ball that define the interaction.

It's the interaction that gives you and the ball reality.

You and the ball aren't really there in the way that you think you are.

You and the ball *seem* to be there. I mean, really *seem* like it.

But because you are both made of particles and forces, and the particles aren't quite there in the way we think they are, and we're not really sure what forces are, except that they are the *whatever* that interacts with particles to make particles interact with other particles (eventually, your brain, your arm, your hand, and the ball), then ...

Reality is not what you think it is. It's kind of backwards. It's the ...

... interaction ...

that creates reality.

So when 7+billion humans fit into two-thirds of an M&M and it seems like none of us are really here, we are here because quarks interact with gluons, which are the particle

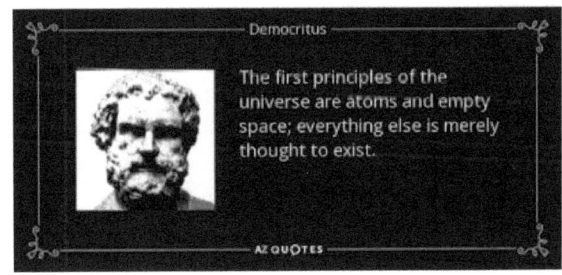

Democritus

The first principles of the universe are atoms and empty space; everything else is merely thought to exist.

AZ QUOTES

Except for atoms, which don't exist either. So close.

that makes the Strong Force work, and the quarks become protons and neutrons, which interact with gluons again to become atomic nuclei, and then the nucleus and the electrons interact via the Electromagnetic Force to become atoms and all of that empty space, and voila!

Here we are.

May the Force be with you.

Because if it isn't, then you aren't, either.

So, where are you then, anyway?

A fine, fine question.

Chapter 22   There's Nothing There – Part 20

We've talked about There. Let's talk about Where. And maybe, When.

Every now and then, someone will ask, so, where did Big Bang happen?, for example. Where's the center of the universe? Where's the starting point?

Back before Copernicus and Galileo and Bruno came along, we thought we had the answer. The earth is the center of the universe. The Church had it all figured out.

It was logical, the Church said. God loves us. He put us on the earth. He must love the earth. We are the relational center of the universe, created by God to be in relationship with him. So the earth must be the physical center of the universe. God rotates around us, so the universe must rotate around the earth.

Well. First, that's neither logical nor necessary. Whether or not you accept the premise that God loves us (and you may not), it does not begin to follow that the earth is the center of the universe. You love

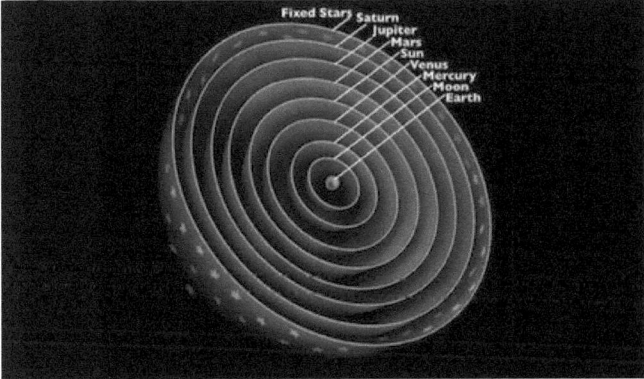

your loved ones (duh), but that does not make them the center of anything except perhaps as a constant drain on your finances.

So, at some cost and courage, Copernicus and Galileo and Bruno fixed all of that, and eventually, everyone got comfortable with the idea that the earth was not the center of the universe, even religious people. Nobody's theology in any of the world's major faiths includes Geocentrism as a plank in the platform.

Which is a fine example showing us that religion and religious people can get past science and accept what seem to be challenges to the faith without losing that faith.

So then we got the Copernican Principle, which says that not only is the earth not the center of the universe, but the earth isn't special, humans aren't special, the universe itself isn't special, nothing at all anywhere is special. Everything is very very ordinary.

As it turns out, the Copernican Principle made the same mistake with science that Geocentrism did with religion. That is, it takes one truth (the earth is not the center of the universe) and then jumps off in an entirely illogical and unconnected direction to say that the earth therefore is not special.

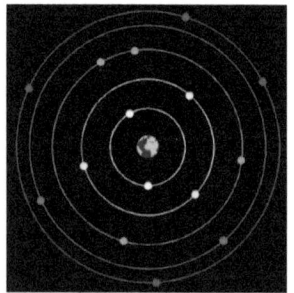

That is, just because your loved ones are not the center of the universe doesn't make them unspecial.

So what the Church said was, humans are special, humans live on the earth, so the earth is the center of the universe.

Here's what science said: the earth is not the center of the universe, so humans aren't special.

What's bad theology going one direction is bad logic going the other direction.

Whether or not humans are special has absolutely positively without a doubt 100% maybe even 1000% (yeah, that's stupid) nothing to do with whether or not the earth is or is not the center of anything.

Everybody (outside of religion) believed that since earth isn't the center, then humans aren't special, though, and still pretty much believes it.

Most scientists and atheists believe that humans aren't special. Neither are the earth, the solar system, the Milky Way, any of the other galaxies, or even the universe itself. Everything is just a big accident produced by the laws of nature, and they don't give a rat's patootie about you or the universe.

See where bad logic gets you? All in a flutter.

But then Big Bang came along and changed the whole game anyway.

Because there was a starting point, it looks like the earth is very special after all.

Here's a headline from Scientific American magazine in February 2016:

Exoplanet Census Suggests Earth Is Special after All

A new tally proposes that roughly 700 quintillion terrestrial exoplanets are likely to exist across the observable universe—most vastly different from Earth.

Here's a short bit from the article: "...*Max Tegmark from the Massachusetts Institute of Technology ... thinks Earth is a colossal violation of the Copernican principle...*"

For us, it's a fine-tuning thing. As we mentioned once or twice.

And, as it happens, the earth is the center of the universe.

Well, the center of its own universe.

Works like this.

Asking the questions, "Where did Big Bang happen? Where's the center of the universe?

If the universe is expanding, isn't it expanding away from one point, a starting point?" comes from not really understanding Big Bang.

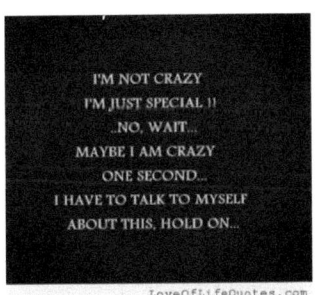

Big Bang was not just the starting point of everything in the universe - planets, stars, galaxies, all of that. It was the starting point of space and time.

Which means that it was the starting point of all the points in the universe.

Which means that all of the points in the universe were once all at the same place. They were all the same point.

Which means that every point in the universe is the point that the universe is expanding away from.

Every point is the starting point.

Every point is the center of the universe. Actually, the center of its own view of the universe, the parts that are visible from each point, which are different from each point.

So the earth is the center of its own universe. So are you, of course. So is your elbow. And the tip of your nose. And all of your loved ones. Duh.

So the first answer to the question Where? is that all of the Wheres are the center of the universe. It's a Big Bang, relativity thing. Like the picture below, everybody is in fact the center of the universe. All of our universes overlap quite a bit, of course.

Now, there's the Quantum Wheres, too.

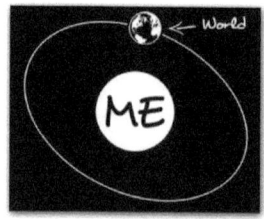

As in, for photons or anything traveling at the SOL, everything is just like it was back at the Beginning, at the Singularity, when all of the points where all at the same place. One point. All of the points in the universe for a photon *right now!* are at the same point.

And all of the Time in the universe happens at the same Time. Everything happens at once. If you happen to be traveling at the SOL. So *right now!* is always *right now!*

Photons are never late for anything, and they never go to the wrong place. Male photons never have to ask for directions, and female photons can take as long as they want to get dressed. If that sounds sexist to you, then just switch them around. Still works.

So all the Wheres and all the Whens are just one Where and One When. Which sounds like Dr. Seuss on steroids.

And even if you're not traveling at the SOL, if you are a quantum particle, you can occupy multiple Wheres and multiple Whens. Particles can be and are in many places at the same time, and sometimes they can even be in many times at the same, uh, at the same, uh, well, OK, it's not at the same time. They can be in many times at the same place. But they can also be in many times in many places, and many places at the same time.

And there's even a Where and a When for you yourself. Well, actually, that's not true. You don't get your own personal Where or When.

Simple question, just for starters. Where are you? Like, right now?

Answer - your Where is relative to all the Wheres that the universe has to offer.

Any answer you give is in relation to somewhere else. And as we all know by now, "relation" is code for "interaction". So you only have a location as you interact with it.

In your room. Your car. Your office. The bathtub. A ski lift.

North America. Earth. The Solar System. The Milky Way. Um, Mars.

Or Nevada. It's hard to tell.

And all of those Wheres are moving. So you are moving, too. Your Where used to be ... *There* ... but now it's ... *Here*. And Here is always moving, changing, interacting with lots of other Heres and Theres.

Consider. Let's say you are in empty space in a spacesuit, just hanging out.

Now. Where are you?

There's nothing around you. No planets, stars, galaxies. Nothing but blackness. And you.

So. Where are you?

If there's nothing else in space (Space-Time, to be more specific), then you have no location.

Now. Here comes some other guy in a spacesuit. Zip! Off he goes. Or she.

So. Who's moving?

You both are only moving relative to each other. If there's nobody there, then nobody's moving. No body. No motion.

Not only do you not have a location, you only have motion relative to other things. Even motion and location are results of interactions.

And since you can't tell how fast you are moving, or that you are moving at all, and since Space and Time change with speed, but only relative to the places you left behind, then they can't change. Whatever that means.

Bottom line at the end of the day in the final analysis when all is said and done, Where and When only exist because of interactions between things made of matter, with mass, and of course, photons.

We should probably talk about Schroedinger's Cat now. Since cats, like everything else, aren't actually There until ...

You look at them. In a quantum sort of way.

Chapter 23   There's Nothing There – Part 21

Now. Schroedinger's Cat is all a part of the Quantum Physics thing, and let's just remind ourselves right up front that 1) Quantum Physics is the Best Science Ever (BSE) which is 2) never wrong and it's always right and had 3) more evidence in support of it

than Any Other Science Ever (AOSE) but is 4) entirely weird and 5) nobody understands it and 6) everybody wishes it would just go away but 7) it explains perfectly how matter and energy work because it is 8) the BSE.

So let's just say right up front that 9) we have a way to maybe fix all of that. 10) Watch this. And 11) you're not going to understand this because 12) #5 up there in the first paragraph. Nobody understands it. So don't feel bad or dumb or ignorant or stupid.

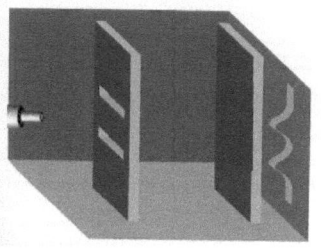

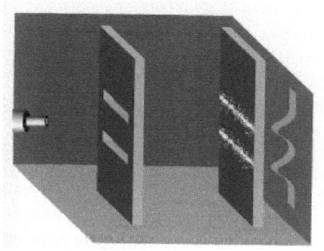

Because I don't understand it either, and here I am, writing about it.

Let's start with the Double Slit Experiment. Here's a picture:

You'll notice that there's a piece of wood with two slits. That's why we call this the Double Slit Experiment.

And there's a flashlight (actually a laser) and another piece of wood.

Now, if the world worked the way you think it ought to, when you turn on the flashlight, you'd get something like the second image:

And sometimes you do. But when the slits are narrow enough, you get something more like this:

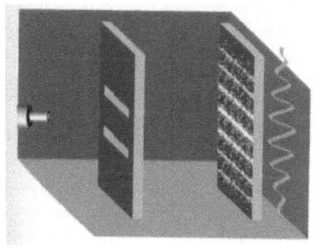

That's because light travels in a wave, and when you shine it through two slits, the wave breaks into two waves and they interfere with each other. So that's called an Interference Pattern. It's not a big deal.

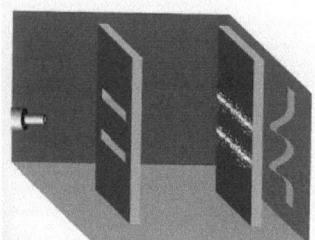

But if you use a laser and change the experiment so that you only send a photon of light through the slits at a time

this is what you ought to get:

That is, some photons go through the top slit and some go through the bottom slit. Just like you'd expect.

Except. That's not what happens. This is what you really get:

Which is an interference pattern.

That can't happen.

The photons are going through one at a time. There's nothing to interfere with.

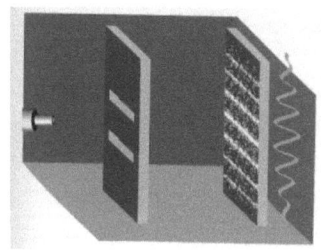

Here's where the weirdness starts.

The photons go through both slits at the same time. That means that each photon is in two places at once. And then each photon ... interferes with itself.

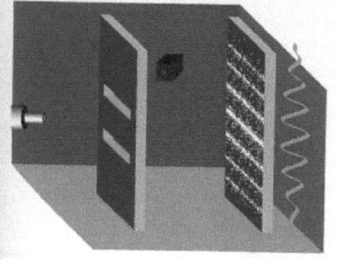

I told you it was weird. You should have listened.

I'm going to say that again really loudly so that you don't miss it.

THEY'RE IN TWO PLACES - TWO!!! - AT THE SAME TIME!

So now we change the experiment by putting a photon detector into the experiment. It ... detects photons. Duh. See the little black box? Photon detector. It's turned off now.

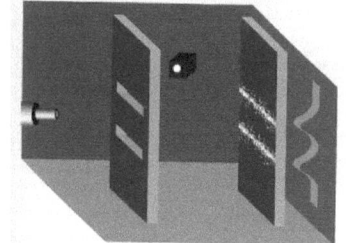

Then we turn it on. And the weirdness gets weirder. Unless you start to understand the universe in a different way.

Because when you turn the photon detector on, the photons start doing what we thought they should have done in the first place. That is, going through either the top or the bottom slit. Not both at the same time.

So what's going on?

Weirdness.

The photons seem somehow to ... know ... that you're looking at them. And they change what they do.

So that you don't miss it ...

THE PHOTONS KNOW THAT YOU'RE LOOKING AT THEM!

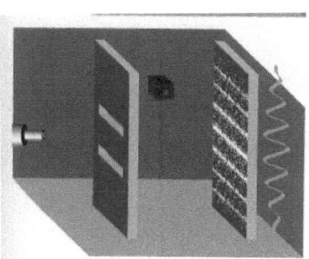

Wait, though. That's so weird. Except that ... what's going on here is ... an Interaction. Between you and the photons. Because "looking" is an interaction.

And the whole universe may be made of just that. Interactions.

So now it all makes sense.

OK, not really. But only because we don't know how to think like that yet.

Then it gets more interesting.

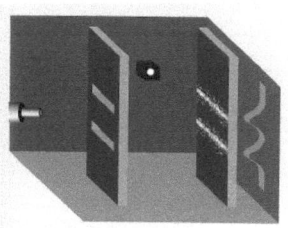

We aim our photon detector towards the middle, between the two pieces of wood. Because then the photons will have already gone through both slits at the same time, and we'll figure out what it is that they are really doing.

And then we turn it on.

And the photons stop going through both slits at the same time again and start doing what they are supposed to do in a non-weirdness universe.

Which means.

That the photons go through both slits at the same time, get to where you are trying to see them, see that you are trying to see them, and then they ...

..go **backwards in time** and change what they did. So that you can't catch them going through both slits at the same time.

Backwards in time. You read it right. That's what they do. Photons, and all little particles, can go backwards in time. You can't, even though you are made of little particles that can.

Once again, loudly.

THEY GO BACKWARDS IN TIME AND CHANGE WHAT THEY DID!

And they do this because you have Interacted with them. By looking at them, or measuring them in some way.

Your Interaction with photons, or any quantum (little tiny) particle, can cause the future to change the present, or the present to change the past, or the future to change the past.

You can't tell it how to do that. It just ... does it. Because of the Interaction.

And remember. These particles aren't really there in the way that you think they are. They are only there as a product of interactions. And the interaction starts with ... You.

Or some random vibration somewhere in the world. Whatever. It's still an interaction.

Still to come, you, interacting with Schroedinger's Cat backwards in time.

Chapter 24   There's Nothing There – Part 22

Here's the problem.

You (and everyone else) think that

1) the world and the universe and nature and everything are just the way that you think they are, and that

2) they would be that way whether or not you are here, because

3) the Copernican Principle. Which says that humans are not special and have no special role to play in the universe. Because

4) we are just one more in a very very long series of accidents happening. And

5) the universe could give a rat's patootie whether or not humans are here. That's the Official Cosmological Term, btw. " Rat's Patootie."

So. Schroedinger's Cat.

Two guys, a Dane and a German, walk into a bar.

OK, not really.

I mean, they may have walked into a bar.
Probably did. But it's no joke.

They were Niels Bohr and Werner Heisenberg.

No, no, not the Heisenberg from *Breaking Bad*. The real one.

He came up with something that has his name on it. It's called the (go ahead, guess) Heisenberg Uncertainty Principle. What, you weren't sure?

There you go.

It says that you can't know both the momentum and location of any particle.

And then they both came up with something that ought to have Bohr's

111

name on it, but that would be dull. Uninteresting. Um. Boring.

Sorry. Bad physics joke.

It's called the Copenhagen Interpretation of Quantum Physics.

(There are 11 different interpretations of Quantum Physics.

That's not really supposed to happen. It just shows what a 1] totally fascinating field of scientific inquiry Quantum physics is, or 2] mess it is. Or 3]. Both.

It's not really a mess, of course. It's perfect, it's never wrong, it's always right, and it's brilliant.

But nobody agrees how it works.

That's not really supposed to happen. Scientists should automatically just *like* something that works all the time. I mean, it ought to be their BFF, except BSF, S being Science.)

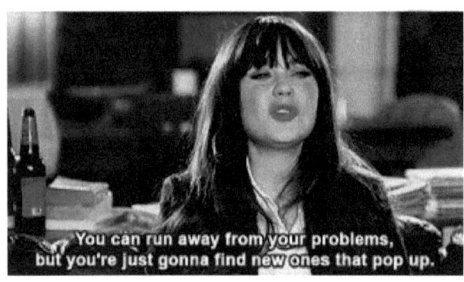

Anyway. I digress. Bohr. Heisenberg. Copenhagen. The Copenhagen Interpretation of QM (Quantum Mechanics which is the same as Quantum Physics and Quantum Theory) says that ...

That a particle has neither a momentum nor a location until you look at it.

That is, it is your interaction with the particle that causes it to, I don't know, exist at all.

So two other guys didn't like that at all. A Wuerttembergian/stateless/Swiss/Austrian/not Austrian/German/not German /American guy and a German/Irish guy went into a bar. Yeah, not really. It was Albert Einstein and Erwin Schroedinger, who was named after a famous cat. Quantum Time being what it is, that might be true.

Anyway. They didn't like the whole observer thing, so they sat down and came up with a *thought* experiment to show why it was all wrong. Of course, Quantum Theory being what it is, they just showed it was all right, instead. Curse you, Quantum Theory. They just might have been in a bar, after all.

So. Here it is. They said, OK, you Copenhageners, what if we take an alpha particle and put it in a box, like that box right there:

This Alpha Particle will eventually decay. Because it's all by itself and not in a big lump of Alpha Particles, we don't know when it will decay and can't predict it:

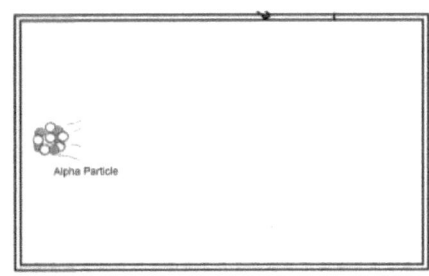

But here's what you're trying to tell us. You're trying to tell us that if we close the box and we're not looking inside, if it's completely cut off from the outside world (sound, smell, taste, touch, and hearing, and any old measuring type device), then ...

The Alpha Particle is in a state of Quantum Uncertainty (like, the Uncertainty Principle). And THAT means that ...

It has both decayed and not decayed. Both at the same time.

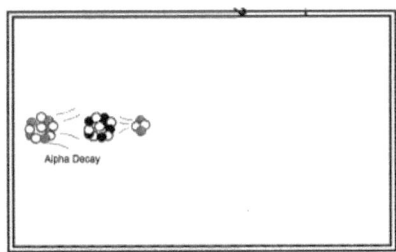

Not just ONE PARTICLE IN THE TWO PLACES AT THE SAME TIME!

But, ONE PARTICLE IN TWO STATES AT THE SAME TIME!

Which is, Einstein and Schroedinger said, ridiculous.

So we're gonna put a bunch of other stuff in the box, all connected to the particle. What, you might ask, could that be? It will be a Geiger counter, a hammer, some poison, and a cat, like this:

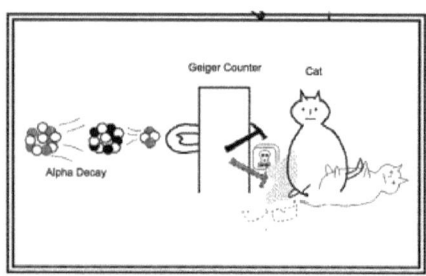

So if the particle decays, the Geiger counter will detect it, let the hammer fall on the poison, which will spill and kill the cat. The cat was Einstein's idea. I don't know why.

So, now, you will be telling us, not only is the particle in two states at the same time, but so is everything else in the box.

And thus ...

The cat will be both dead and alive. At the same time.

THE CAT WILL BE BOTH DEAD AND ALIVE!

And that's ridiculous. Said Albert and Erwin to Niels and Werner.

And Niels and Werner said, ah, we don't wanna make you guys feel all bad and offended, but, um,

That's the way it is. And they were right. And Einstein and Schroedinger were wrong.

And then, what's more, is you open the box and look inside.

And then what happens is that you, um, ah, cause reality to come into being. As in, the particle was in two states at the same time, which means that reality didn't exist yet, and then you

interacted

with it, and voila! Reality!

It's like this. When you look inside or make a measurement or whatever, you do what's called ...

... "collapsing the probability wave function" ...

which means (Copenhagenishly) that, where before the cat was both alive and dead, or neither alive nor dead, or whatever, now ...

... the cat is one or the other. But not both.

And this happens only when you look inside the box. Before, dead and alive. After, dead or alive. You did this. You yourself.

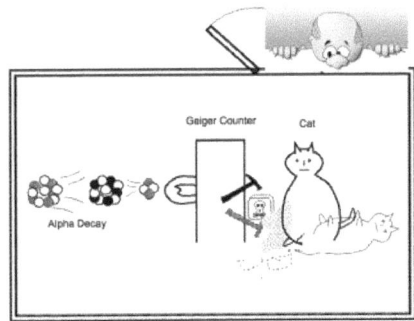

You need to know that you don't get to choose whether the cat is alive or dead. The particle does that for you. Because this is a story about a particle.

As in. The particle notices (whatever the heck THAT means) that you have looked in the box, and sort of runs through its options.

Option one - the particle decays sometime in the future.

Option two - the particle decays right now, when you look in the box.

Option three - are you sure you want me to tell you? Your life will be much much easier if you just stick with Options One and Two.

Really? OK, then. No pain, no reality.

Option three - the particle decays sometime in the past. Like, yesterday. Or last week.

But it doesn't decay last week until you look in the box.

Which means that the cat isn't dead until you look in the box.

Which means that when you look in the box, THEN the cat will have been dead for a week.

But not until you look in the box. Which means that 1) the wave function collapsed backwards in time and therefore 2) your interaction with the particle *right now* affected things that happened *in the past*.

Of course, technically they didn't happen in the past until your interaction. That's just confusing.

Interaction killed the cat. Of course, you were curious about what would happen if you looked in the box, so then, of course, like a force of nature ...

... curiosity killed the cat.

Ha. Hysterical.

Anyway. What THIS means is that ...

... without interaction, there is no reality in the universe.

Do I need to repeat that really loudly so that you'll get it?

I didn't think so.

And reality can come into being ... backwards in time.

I think I have a headache.

Yesterday. Ha. Hysterical.

Chapter 25   There's Nothing There – Part 23

Now. Although nobody is all that crazy about reality not existing unless some sort of observation or measurement is made (apparently by some sort of intelligent being, like a human or ET, but not by a really smart animal, like a bonobo)(You need to go look up "bonobo" right now), nobody can get away from it.

The best that anyone can do is to try to figure out what happens AFTER the observation is made. But nobody has been able to say that the observation is NOT the thing. The Thing. The Thing that makes reality come into being.

And nobody can really define what an observation actually is.

Or what the observer actually is. Or who.

Now. I'm fully aware that all of this sounds insane. Which is part of the problem. It just sounds crazy. OK, that's ALL of the problem. It's lunacy. But it's the way the universe works.

The Copenhagen Interpretation says that all possible realities exist together (cat both alive and dead) until the observation is made, and then all the possible realities go away except for one. Live cat universe OR dead cat universe.

The Many Worlds Interpretation says that when the observation is made, the universe splits into two universes. Live cat universe AND dead cat universe.

That would mean that there are many many many universes. Except, of course, there aren't. At least none that we have any evidence for. This one, of course. Ours. But not a single other one.

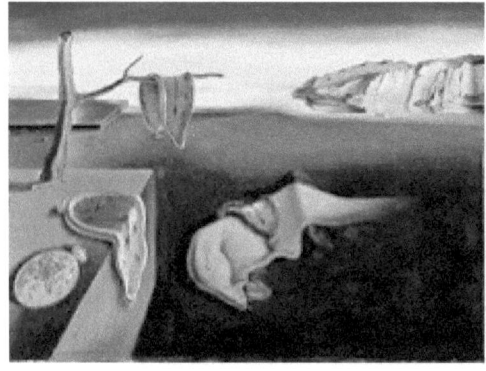
Something like two out of three physicists accept the Copenhagen. Something like one out of nine accept Many Worlds. That's about three out of four of all physicists. The other one physicist likes any of the other nine interpretations. So they aren't too popular.

Really what's going on is that a lot of very smart people are hoping that there's (and this is what they call it, so use your deep professional American Idol voice) ...

A DEEPER REALITY THAT MAKES SENSE.

Albert Einstein wanted this. Stephen Hawking wants it. Roger Penrose. Lots of folks. Most everyone.

And here's what's interesting about that.

It's that ... they don't like it because it doesn't make sense to them, even though it's perfect and it's always right and it's never wrong.

That is to say that they fully believe that the universe should be understandable to us, the beings that live in it. Since ET hasn't shown up yet.

Even though Neil deGrasse Tyson says quite firmly, "The universe is under no obligation to make sense to you," he really does want it to make sense to him, even though it's under no obligation to do so. That's what he himself said to him himself, along with all the rest of us ourselves.

We could note here that we are the only creatures on earth, and maybe the only creatures in the universe, that gives a rat's patootie about understanding the universe.

All the other creatures just eat, sleep, excrete, procreate, and try to keep daddy from eating the babies.

We (firmly convinced that we are not special)(which we figured out all by ourselves)(based on fairly bad logic, as it happens) spend a good amount of our time trying to understand things:

Physicists - the universe.

Other scientists - smaller parts of the universe.

Psychologists - why we are such whackadoodles.

Historians - why we are such whackadoodles.

Marketing people - why we won't buy their crap, and how to make us buy their crap.

People - all the other people.

Animals don't do this. Plants don't do this. Galaxies don't do this. Even particles, which apparently are smarter than we are, don't do this. Particles are just flipping coins to figure out what they are going to do, and waiting to see if we are going to try to catch them doing it. But particles are not trying to understand particles.

We're the only ones.

And if we reach a place where we can't understand things (like space-time, black holes, Big Bang, quantum things, relativistic things, bendy time and bendy space, cats both dead and alive until we look at them, particles going backwards and forwards in time, stuff like that), then instead of just saying, wow, we're just not gonna get that, are we?, we say, dang, the universe must be broken, cause it's way weirder than we thought it was going to be.

And what the universe is saying is, hey, it's YOU guys who are weird. The rest of it? That's normal. That's the way it is. You're like little babies afraid of the dark or the bathtub or bunnies or green veg. Or, in my daughter's case, ET, spiders and laundry lint.

So somehow, we who are not special are supposed to be the only creatures in Space-Time who are special enough to be able to figure everything out, how it all works and makes sense to us who are like tiny little rancid carbuncles on the right butt cheek of the solar system.

The universe does throw us a bone. It lets us find and understand the math that describes it all. The math of relativity, of the general theory, of

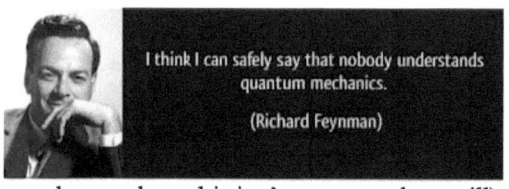

quantum theory and the Standard Model of Particle physics, of Chaos and Complexity, even String Theory (OK, we don't quite get the math and it isn't true yet, but still) and Loop Quantum Gravity (ditto) and Inflation (ditto ditto) and Dark Energy and Dark Matter (which EXIST but we CAN'T FIND THEM!!!).

But the math, the cruel, bitter heartless math, tells us a story that we cannot understand. And will not. But we want

so

badly

to understand it.

It's kind of adorable.

Because we, cute little Minions that we are, make this assumption that has no support.

And it is, we can Understand Everything.

We who can't see the backs of our own heads, we who live in a universe that is maybe 10^{26} times bigger than we can actually see, we who look out at the night sky, see 9110 stars (which is all that we can see with our eyeballs) and assume that we are Seeing It All, we for whom 95% of the universe is Dark Energy and Dark Matter and we have no idea what they are, we who cannot go backwards in time or tunnel instantaneously across vast regions of space or be in 2 places at once, though all of our particles can, we who travel to the Moon and think we have been in space, we who dream of colonizing Mars, which is a lot like dreaming of colonizing that spot in the sand right next to us on the beach, we who are bound to air and water and food and gravity and light and cannot live without any of them,

We want to Understand it All.

Adorable. Just ... adorable.

Chapter 26   There's Nothing There – The End

Now I'm going to say something that will irritate some of you.

Here it comes.

Nobody is smart enough to be an atheist.

Go ahead. Write me nasty letters. Be a troll. Find your inner troll and let it out to stomp me into troll-stomping oblivion.

Wait, though.

I'm going to say something that will irritate the rest of you.

Here it comes.

Nobody is smart enough to be not an atheist.

Go ahead. Write me letters that pray for me and consign me to the darkest depths of hottest hell. Send the demons after me. Find your inner demon and let it out to stomp me into demon-stomping oblivion.

Wait, though.

Let's think about it for a minute.

'Cause what we're really saying is that belief is not really a matter of intelligence.

Now THAT will open up dozens of worm cans. Cans of worms. Whatever.

Belief is also not a matter of information. As in, the more you know, the less you believe.

There you go. More worms in more cans.

Let's do the easy part first.

As in. No matter how smart you are, there's somebody smarter who has different beliefs about The Old One (as Einstein called God).

Bingo. Einstein believed in The Old One, and he is officially smarter than nearly everybody except for maybe Isaac Newton (who believed) and maybe da Vinci (who believed, even though he painted naked people)(that is, he painted pictures of naked people, not, he painted people who were naked. I mean, he did, but not on their actual nakedness.)(I think he may have opened a tattoo parlor in his later years.)(Some of that was not strictly speaking true.)

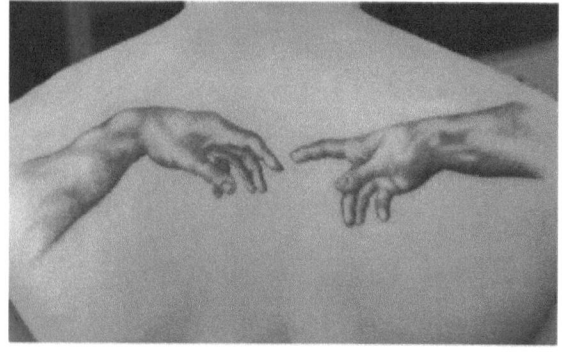

So. But. Einstein was not a practicing religious person. Not as a Jew. Not as a Christian. Not as anything.

So he's the metaphor. If you want to disbelieve, you can't point to Einstein as your example, and if you want to believe, you can't point to Einstein.

And you are not smarter than Einstein.

Of course, you're not smarter than Newton or da Vinci, either. Sorry. No offense.

For most of you, there's a lot of people that are smarter than you.

Some of them believe. Some of them don't.

So it's not really an intelligence thing.

How about information then?

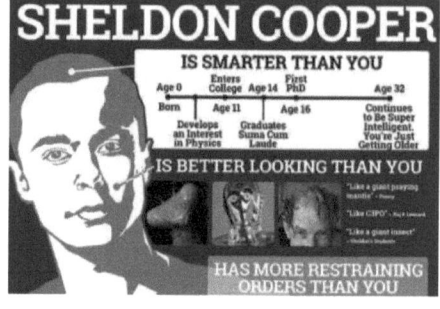

Well. The problem with information is that it changes all the time.

And when information changes, evidence changes.

And when evidence changes, facts change.

And when facts change, science changes.

'Cause science is about facts, facts come from evidence, and evidence changes all the time.

We used to think, for example, that the universe made sense. That it was predictable. That the laws of physics caused everything to happen. That we didn't need God to explain anything. That the

universe was infinitely large and old and everything had always been here, including the laws of physics.

But we didn't know what the laws of physics were yet.

We thought there was just gravity. And as it turns out, we didn't even understand gravity.

And then when Einstein explained gravity to us, all of sudden, the universe wasn't infinite at all.

And the laws of physics hadn't always been here. Which means that something that wasn't the laws of physics caused the laws of physics.

And boy, were they weird.

So everything changed. All of science changed. All the evidence changed. All the information changed.

And it might change again.

No. That's wrong.

It will change again.

So you can't base your beliefs on information that can and will change.

Sometimes it'll make it seem like there is no God, no creator.

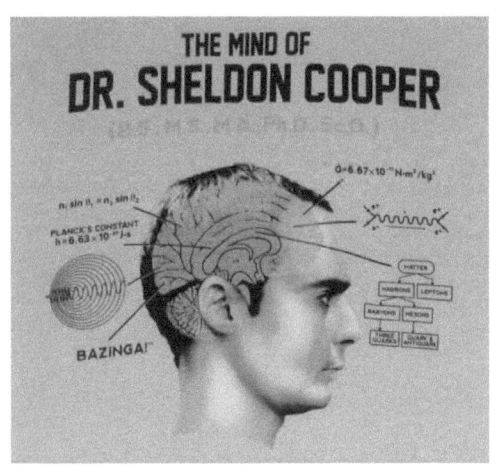

Sometimes it'll make it seem like there might be one. Or there must be one.

And then. It'll change again.

So even if you are Ed Witten. He's the smartest guy alive at the moment. Or Stephen Hawking. Or Marilyn Vos Savant (she has the highest IQ on the planet). Or my friend Chip Diggins (his IQ is scary high.) Or Richard Dawkins (he's only kinda smart.) Or Father Andrew Pinsent at Oxford (my very very smart friend Simon says Fr. Andrew is smarter than Simon is.)(Simon got his PhD in particle physics from Oxford and did his dissertation at CERN, so as we say in Texas, he's pretty dad-gummed smart.) Or Rev. John Polkinghorne. Or Paul Davies. Or Fred Hoyle. Or Neil deGrasse Tyson. Or Stephen Weinberg. Or, frankly, anybody. Even Sheldon Cooper.

It doesn't matter how smart you are, or how much you know. Because there's someone smarter who believes differently than you do, and what you know, will change. And then it will change again.

It's not about intelligence. It's not about information. It's about something else.

And what, you may ask, might that be?

Now that is a stellar question.

Chapter 27　　　There's Something There - Part 1?

It's Part 1 because there might be other parts, but I don't really know yet. We'll just have to see.

Anyway. Since we're being provocative ...

What you believe about _____ (go ahead, fill in the blank) is irrelevant.

OK, I'll fill in the blank for you.

What you believe about <u>God</u> is irrelevant.

And ...

What you believe about <u>the Universe</u> is irrelevant.

Why, you may ask, am I making such obnoxious statements? Why, you may ask, am I being such a jerk?

Here, let me add some emphasis, see if that helps.

What you *believe* about <u>God</u> is irrelevant.

What you *believe* about <u>the Universe</u> is irrelevant.

Let me explain. We'll start with the Universe.

Here's the Thing. The Universe is the way that it is regardless of what you believe about it. It doesn't care what you believe. As we already said that Neal deGrasse Tyson said, the universe is under no obligation to make sense to you. Or to anybody.

If you think it doesn't make sense, that's not the Universe's fault or problem. It's your problem.

The picture below is not the universe exploding. It's just your garden-variety star exploding.

Actually, not exploding. Collapsing and bouncing.

The nothing didn't explode either. It just expanded really really fast.

You can believe that it's infinitely large, or not.

You can believe that it had a starting point, or not.

You can believe that it's 6000 years old, or 13.8 billion. Or infinitely old.

All that you, and we, and everybody has to go on is the evidence that we have been able to uncover, and will presumably continue to uncover, and we interpret that evidence from our tiny perch in time and space as best we can.

And as we have seen, the evidence changes, and so our understanding of the Universe changes, and presumably it will continue to do so, whether we like it or not.

The Universe is indifferent to your feelings about it. It's like that cheerleader that I asked out four times in college. Her name was Monte Vaughn. I still remember her name. She was ... indifferent. And remained indifferent. She did not go out with me. She does not remember my name. She may not have actually been a cheerleader, either. But it's a better story that way.

If Monte Vaughn had shown this much emotion when I asked her out, well, a win.

I just checked. She was Homecoming Queen. OMG. No wonder.

The Universe does not remember your name. It is the way that it is, and it will continue to be that way (that is, dynamic and constantly changing) regardless of your thoughts, opinions or beliefs about it.

If it's helpful to rename the Universe "Monte Vaughn", go right ahead. It is unimaginably beautiful. And it is not interested in going out with you.

Now. God.

God is something like the Universe.

Either God exists, or he does not. What you believe about that is irrelevant.

That's harsh, but true. But you will not eliminate God with your lack of belief, if he exists. Nor will you cause him to exist via your beliefs, if he does not.

Either he does, or he does not.

(I'm using "he" because constantly having to say "he, she, it or they" is awkward and irritating and so I'm not really saying God is a man or male, but that he is surely more than an "it". "They" would work, but it will irritate the non-Trinity folks and yadda yadda yadda. Live with it.)

Either he does, or he does not.

Your opinion about that is immaterial.

Now isn't that a curious thing? It causes one to wonder, ok, NOW what am I supposed to do?

Well. We could talk about evidence for, let's say, the universe first. But we will quickly run into theories that say, as we have said in other earlier blogs, 1) the universe is a hologram (and therefore not really here) or 2) the universe is a computer simulation (and therefore not really here.)

Plus, since we don't know what either Dark Matter or Dark Energy are, 95% of the universe is a complete mystery to us.

Plus, the universe might be 10^{26}th times bigger than the part that we can see, and we'll never see any of the rest of it and will never know anything about it,

Or, it might be infinitely big, which is a lot bigger than just 10^{26}th times bigger, and so the part that we do know something about is infinitely smaller than the rest of it, and (just do the math) is therefore not really here,

And that's THE OBSERVABLE UNIVERSE, of which you are not even a measurable part of, either. Well, measurable maybe. But very close to infinitely insignificant,

And the part of the universe that we can see, we will never ever be able to go to any parts of it, apart from the part that we are in. Like, even getting to Neptune seems like a stretch. Frankly, even Mars seems like a

stretch. I think you could put all the people who've been to the moon in an egg carton. OK, a large egg carton. One that would fit people. Or a roller coaster that seats 12. Like that one. That many people have been on the moon.

And we will never ever *know* for sure where the universe came from, or whether it is unique and there are lots of other universes, or maybe just a couple or four, a dozen maybe, a gross of universes.

Cleverness

a good thing?

(If only I knew)

(OK, to be honest, you can't actually be "very close to infinitely insignificant", but if you could, we would be.)

So *your* personal evidence for the existence of the universe is anecdotal and highly suspect. I would not choose to believe in the universe based solely upon *your* personal experience.

And I don't give a rat's patootie who you are, because even you, you universe scientists, you cosmologists and astronomers, even you have only seen or experienced a tiny tiny fraction that (if the universe is infinitely big) is infinitely smaller than the actual universe itself, and you don't even have much of a clue what it is, either. And you keep changing the story.

No offense. It's kinda hard. I get that. That's actually the point.

I should say, BTW, that I do in fact believe in the actual universe.

Pay attention. This is the subtle and clever part.

1) I understand the universe as well as I understand it because the cosmologists and astronomers have gone out of their way to explain it to me, and 2) I understand that my understanding of the universe will change with further information, and 3) it might change quite radically in quite unexpected sorts of ways, but 4) I have personal experience with the universe. That is, 5) evidence has been presented to me, and 6) I have personal evidence as well.

The universe exists for me because I have some experience with it.

It is not the same experience that anyone else has had, though there is a lot of overlap, so it is both unique and universal at the same time.

Unique and universal. How...

Intriguing.

Chapter 28   There's Something There - Part 2

It apparently happened in S. Korea and became a boy band.

Nothing right and good about that.

I read the other day that the Big Bang didn't happen in space.

Which makes sense, since there was no space yet for it to happen in.

Still. That's hard to get.

So there was no Where that it happened in.

But there was a When.

Big Bang happened in a When, but not a Where.

It happened in time, but not in space.

So if you were to ask, sweetly, naively, where did Big Bang happen, thinking, sweetly, naively, that there had to be a point where it all started, the answer is ...

There is a point where it all started. But it's a point in time, not in space.

So the evidence in science has taken us to a place in nature where our brains cannot go.

It's on Thursdays at 9. See. It happens in time. Even if you count syndication.

Big Bang happened in time. Not in space.

Do you want a list of all the other places that the evidence has taken us that our brains cannot go? Just go back and read some of the earlier chapters. Any of them.

131

So now, let's talk about God.

Pay attention. This is the subtle and clever part.

1) I understand God as well as I understand him because the theologians have gone out of their way to explain him to me, and 2) I understand that my understanding of God will change with further information, and 3) it might change quite radically in quite unexpected sorts of ways, but 4) I have personal experience with God. That is, 5) evidence has been presented to me, and 6) I have personal evidence as well.

God exists for me because I have some experience with him.

It is not the same experience that anyone else has had, though there is a lot of overlap, so it is both unique and universal at the same time.

Unique and universal. How...

Intriguing.

Now, I could go through the evidence for God that might exist in nature, but I've kinda already done

This would be the Ironic God. Ha. So funny.

that for like almost an infinite number of earlier chapters (OK, 27, but still. It was a lot.), and here's what's happened. I predict.

Yes, it's a chaotic, unpredictable universe, but only mostly, so I'm gonna live on the edge and make a prediction.

Those who already believe in God were like, YES!, and, AWESOME!, except for those who don't like Big Bang, but you probably stopped reading a long long time ago, which was foolish. No offense. OK, maybe a little. But you're probably not reading anymore anyway. No offense.

And those who don't believe in God were like, oh, seriously, all of this crapload of pseudo-scientific religio-babble AGAIN?!?! When are you religious morons

gonna catch a clue that there is NO EVIDENCE for God to be found in nature?!?!

You should maybe dial it down a little. Fewer all caps. By everyone.

Anyway. Here's the thing.

We're all looking at the same evidence. But we are each reaching a different conclusion.

Same evidence. Different beliefs.

So it's not really about the evidence. I mean, it helps. But when one group looks at Big Bang and says, whoa, looks like God to me, and another group looks at the same Big Bang and says, wow, looks like a

pretty cool accident of nature to me (and a third group looks at the same Big Bang and says, don't even THINK of trying to tell me that the God that I believe in would have used Big Bang to create the universe.)(Sorry about the all-caps.), then ...

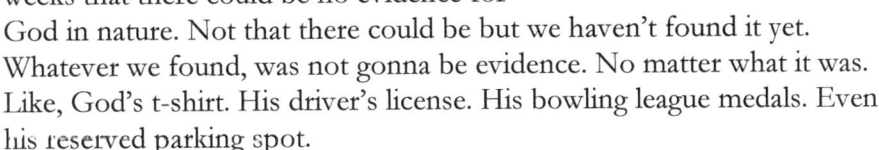

It's not really about the evidence.

I had a guy tell me once over and over for weeks that there could be no evidence for God in nature. Not that there could be but we haven't found it yet. Whatever we found, was not gonna be evidence. No matter what it was. Like, God's t-shirt. His driver's license. His bowling league medals. Even his reserved parking spot.

When I mentioned that there were a host of notable scientists who

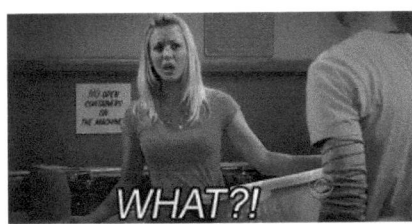

believed that there was some evidence for God, this guy just said, well, they're all wrong.

When I told him that all of the scientists were a lot smarter than he was, he said, no, they're not.

Gotta admire his commitment to his beliefs. Takes some backbone to say that evidence is not evidence and is never going to be evidence and that on the basis of this assumption, and this alone, that he is smarter than all the scientists except for the ones that agree with him.

Faith. That's what it takes. Faith.

Of course, what he believes about God or science or evidence or nature or the universe is, as we've said, irrelevant.

God exists, or he does not.

Even if there is no evidence, God still might exist. Even with lots and lots of evidence, he might not.

So how in God's name are we supposed to figure it out? So to speak.

Here's a question. What would evidence for God's existence look like?

Here's what I think.

Here's a proposal. Here's what it takes. Here's how to start.

It's like, an equation.

It's evidence plus experience plus faith.

Plus one more entirely interesting, controversial, subjective thing that depends entirely on one's interpretation. Kinda like quantum mechanics. Lots of interpretations. Only one is right. We just don't know which one.

Anyway. One more thing.

It's evidence plus experience plus faith plus revelation.

Oh, yeah. It's an equation. It's gotta "equals" something.

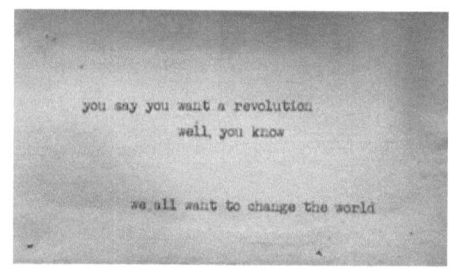

No, not a revolution. A revelation.

Same thing, maybe. Evidence plus experience plus faith plus revelation equals maybe there is a God.

We might start with that. Revelation.

If God created time and space, then God is outside of time and space.

And therefore we can't find him unless he wants to be found.

That is, you can't go on a search to find God. Well, you can, but good luck with that.

It's like the Ultimate Hide-&-Seek game. Ha ha, you'll never find me because I'm outside the universe.

Here's what Big Bang cosmology says: Space and time ... came into existence in a tiny tiny tiny fraction of a second, everything coming from nothing, out of nothing, caused by nothing.

And God. Has to be in the nothing.

And we can't go there.

He. Has to come here.

Ah, shoot. I went and gave it all away. Spoiler alert. I HATE it when that happens.

Fewer all caps. Gotta work on that.

And here's the really devious part. Let's just assume for fun that God is in the fifth dimension. No, not the old rock group, though God could clearly be black and groovy. Or hip. Or whatever.

We live in 4 dimensions, 3 of space, 1 of time.

So let's go with God lives in a 4th spatial dimension. I don't know if he does, but I don't know that he doesn't, so it'll work.

Now. First, consider that there might be a world that exists in 3 dimensions, 2 of space and 1 of time.

Flatland. Like a table top. The people are flat people and they only know of 2 dimensions. This way, and that way, but not up or down.

And you live in 3 spatial dimensions, just like you actually do right now anyway.

Now. How close can you get to the Flatlanders without them seeing you? Because you are in the 3rd dimension, and they are only in the first two.

Take a flat surface right now and see how close you can move your hand without it touching.

Very very very close.

So if God lives in a 4th spatial dimension, how close can he get to you without you knowing about it?

Very very very close.

And you would never know that he was there. Unless he wanted you to know. And that would only happen if he ...

... touched your version of Flatland.

Now I got goosebumps.

Chapter 29   There's Something There – Part 3

Interaction.

That is, the only way for you to find God is ...

For God to find you.

And how, do you suppose, would he do that?

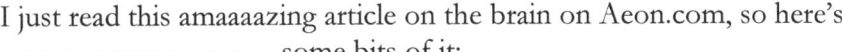

By *interacting* with the universe.

Didn't see that coming, didja?

I just read this amaaaazing article on the brain on Aeon.com, so here's some bits of it:

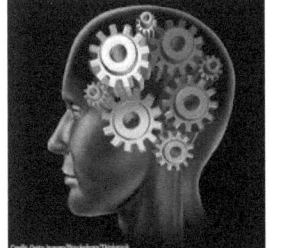

"A few cognitive scientists – notably Anthony Chemero of the University of Cincinnati, the author of Radical Embodied Cognitive Science (2009) – now completely reject the view that the human brain works like a computer.

"The mainstream view is that we, like computers, make sense of the world by performing computations on mental representations of it, but Chemero and others describe another way of understanding intelligent behaviour – as a *direct interaction* between organisms and their world.

"... if we had the ability to take a snapshot of all of the brain's 86 billion neurons and then to simulate the state of those neurons in a computer, that vast pattern would mean nothing outside the body of the brain that produced it ... Whereas computers do store exact copies of data – copies that can persist unchanged for long periods of time, even if the power has been turned off – the brain maintains our intellect only as long

Neural circuits in the brain.
Looks very much like a fractal.

as it remains alive. There is no on-off switch. Either the brain keeps functioning, or we disappear.

"What's more, as the neurobiologist Steven Rose pointed out in *The Future of the Brain* (2005), a snapshot of the brain's current state might also be meaningless unless we knew the *entire life history* of that brain's owner – perhaps even about the *social context* in which he or she was raised."

Which is to say (I'm back, btw. This is me now.) that each of our brains exists as a *direct interaction* with the life we have each lived, each brain unique as each life is unique.

Pardon me for a second. I'm going to make that really really large so that you don't miss it.

Direct interaction

Here's what humans do that is so cute. We look for the rules and regulations that make everything happen, and say to ourselves, wow, that's why everything happens.

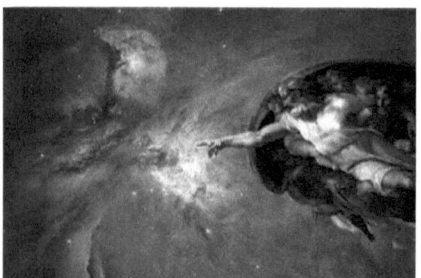

Rules. Regulations. Laws of physics.

Here's what's curious about that.

Science has decided that religion is ridiculous.

But both science and religion make the same mistake.

Religion ... looks for rules for behavior. Human behavior.

Science ... looks for rules of behavior. Human behavior, ultimately, along with the behavior of everything else in the universe.

Science finds its rules in the patterns of mathematics, and then in the patterns of everything. Or vice versa. Sometimes the math tells us what is happening (the Special and General Theories, Quantum Mechanics), sometimes we figure out the math so we can describe what is happening (Newton's Calculus).

But once we had the rules figured out, well, dang it, they stopped making sense about 117+ years ago, in 1900, when Quantum Mechanics came on the scene, and then again in 1905 and 1915 when Special and General Relativity arrived, and then in 1961 Chaos Theory didn't help at all, and when Complexity Theory started to mess with Evolution later in the 1960s, well, that was not a happy moment.

The Rules, the Laws of Physics, sort of stopped helping us understand the way things worked. Instead, they didn't help at all.

Religion kinda has the same problem.

But it's kinda the same problem because it's kinda the same source.

Humans.

We want things to make sense.

So. Let's say some guy comes along. Let's call him Isa. Or Joshua. Or Yeshua. Pick your language base.

And he says, alright, jerks, stop being jerks. Start being nice.

And the whole God thing? You really should pay attention.

So what do religious people do?

Instead of just, you know, being nice, they ask, what does it really *mean* to be nice? I mean, do we have to be nice to *everybody*? Or can we pick and choose?

And what does being nice really *mean*?

So then they made up some rules.

And before long, the religions all became to be about following the rules instead of, you know, being nice.

Buddhism does this. So does Hinduism. Judaism. Shintoism. Jainism. Islam. Paganism. Satanism. Witchcraft. Animism. All the -isms do it.

Christianity, too.

Not an -ism grammatically, but still an -ism. And it does it, too.

Somehow they all start with the interaction between God (or gods)(or no gods)(depending) and humans, and then, somehow, along the way,

they all become about the rules.

And here is what science is telling religion.

It's not about the rules.

It's about the interaction.

If the universe is all about interaction,

then reality is all about interaction,

and the rules of nature come from interaction between particles and forces, and energy and mass and space-time, and observation and reality,

then, well, religion ... is all about interaction. The interaction between us, and God.

So now we gotta figure out what the heck *that* means.

Chapter 30   There's Something There – Part 4

So we think that science is all about the rules, and we think that religion is all about the rules, but

science is turning out to be all about the interactions, and so

religion might be all about the interactions, too.

Not the rules.

Here's what the rules do in science. Oddly enough, they inhibit understanding.

Well, for the most part, they help us understand, but there're those critical places where they don't help anymore.

The Quantum places. The places where folks like the eminent John Wheeler says, if you think you understand quantum mechanics, you don't understand quantum mechanics. And the even eminenter Richard Feynman, who said that nobody understands quantum mechanics.

We can do the math and make the predictions and they always come out right and they're never wrong, but

we don't understand anything about it at all.

Frankly, even relativity doesn't lend itself to understanding, either.

Here's a bit from Roger Penrose:

"Quantum reality is strange in many ways. Individual quantum particles can, at one time, be in two different places - or three, or four, or spread out throughout some region, perhaps wiggling around like a wave.

"Indeed, the 'reality' that quantum theory seems to be telling us to believe in is so far removed from what we are used to that many quantum theorists would tell us to abandon the very notion of reality when considering phenomena at the scale of particles, atoms or even molecules.

"This seems rather hard to take, especially when we are also told that quantum behaviour rules all phenomena, and that even large-scale objects, being built from quantum ingredients, are themselves subject to the same quantum rules.

"Where does quantum non-reality leave off and the physical reality that we actually seem to experience begin to take over?

"Present-day quantum theory has no satisfactory answer to this question.

"My own viewpoint concerning this - and there are many other viewpoints - is that present-day quantum theory is not quite right, and that as the objects under consideration get more massive then the principles of Einstein's general relativity begin to clash with those of quantum mechanics, and a notion of reality that is more in accordance with our experiences will begin to emerge.

"The reader should be warned, however: quantum mechanics as it stands has no accepted observational evidence against it, and all such modifications remain speculative.

"Moreover, even general relativity, involving as it does the idea of a curved space-time, itself diverges from the notions of reality we are used to.

"Whether we look at the universe at the quantum scale or across the vast distances over which the effects of general relativity become clear, then, the common-sense reality of chairs, tables and other material things would seem to dissolve away, to be replaced by a deeper reality inhabiting the world of mathematics."

Reality as science has discovered it is really nothing like the reality that we really think reality really is. Really.

And it gets worse, of course. From *Quanta Magazine*:

"On the other side are quantum physicists, marveling at the strange fact that quantum systems don't seem to be definite objects localized in space until we come along to observe them — whether we are conscious humans or inanimate measuring devices.

"Experiment after experiment has shown — defying common sense — that if we assume that the particles that make up ordinary objects have an objective, observer-independent existence, we get the wrong answers.

"The central lesson of quantum physics is clear: There are no public objects sitting out there in some preexisting space. As the physicist John Wheeler put it, 'Useful as it is under ordinary circumstances to say that the world exists 'out there' independent of us, that view can no longer be upheld.'"

And now you want to complain about God. That is, God the way that you think you understand God.

And you don't even understand chairs.

Which aren't there, anyway, unless you are there to sit on them. Apparently and not quite completely metaphorically.

So you might say, evidence for God has to look like the kind of evidence that you're used to seeing. You want *real* evidence, not quantum relativistic evidence.

Of course, the only real evidence is quantum relativistic evidence, and the "real" evidence that you want is, from your perspective, delusional. That is, Newtonian and "real" only as a tiny splinter bit of the illusion created by quantum relativistic evidence.

But, OK. Let's play that game.

If science is ultimately and finally and completely about the interactions (which is our theory at this point),

and if God exists (which is our theory at this point) ...

then God, who clearly put the whole science schtick together (that's just if a=b and b=c, then a=c) ...

might have a certain commitment to and/or interest in and/or a participatory role to play in ...

Interactions.

Since there doesn't seem to be anything at all but interactions in the universe.

And thus, a revelation seems to be in order.

Because a revelation is an interaction, and interactions are all that there are.

So that would be ... evidence. If there was in fact an interaction that we might say could possibly be in some sense maybe perhaps a revelation.

But you don't want a quantum revelation, like, say, the need for an observer outside of space and time to make an observation and collapse the wave function of the Singularity into Big Bang and hence into the universe itself.

And you may want proof. I should write that as "proof". You are tempted to demand "proof".

If you'll review previous chapters, you might find somewhere that you never ever in science (or religion or on TV crime shows) get "proof".

You get "evidence". "Proof" does not exist. Anywhere at any time. Ever. The only way you can actually use the word "proof" in a sentence is to say you never get any.

Everything that we think might be proven might be disproven when new evidence comes along.

As new evidence tends to do.

So there is no proof of Big Bang. There is only evidence. It's pretty impressive evidence, but you never know. Science didn't use to believe in Big Bang because how ridiculous is THAT! The universe had a starting point. That's absurd. Everybody knows it.

There you go. Everybody, which is to say, everybody in science, and I mean *everybody, all* the scientists were wrong.

And so there never will be any proof that God does or does not exist. Evidence, maybe. Proof? Never.

Unless he, like, shows up. That would do it.

But since he has not, then clearly 1) he does not exist at all or 2) he's got another idea.

And that plan might be all about evidence, and experience, and revelation, and finally, faith.

The plan might be all about faith.

Maybe. If there is a God. And a plan.

Chapter 31   There's Something There – Part 5

Now when religious folk start going on and on about revelations, the revelations always seem to come with rules.

And those rules seem to be weirdly focused on what women are supposed to wear.

Plus, what we are or are not supposed to eat.

And a lot of other stuff that seems to be about what to do and when to do it, or what not to do and when not to do it.

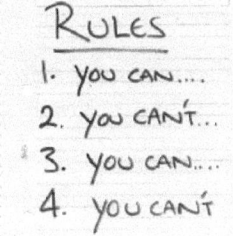

And then we use these rules to make up other rules about who's in, and who's out.

Who belongs to our little group, and who does not.

As though the God of the universe is sitting up there in his 5th dimensional whatever it is spending all of his time being really irritated about us not wearing the right clothing or not getting the right haircuts or not eating food that has been prepared in the exact and precise way that for some reason he made up for no reason at all (eat cows! and chickens! don't eat pigs! or goats! or lobsters! and what the heck is kale?!?! I didn't make that! It's not for *eating*! Where did that awful stuff come from?!?!) and playing Game of Thrones with us (OK, Protestants and Catholics - 1, 2, 3, kill each other!)(OK, Sunnis and Shi'ites - 1, 2, 3, kill each other! And when you finish that, kill all the Jews, even though I like them a lot! And all the westerners. I'm not sure why, but what the heck!), and then sending down songs that all kinda sound alike, and then giving all kinds of strange stage directions (You people, raise your hands

and sway back and forth and close your eyes and make weird noises! And play with snakes! (Really? Snakes?) And you guys, pray the same prayer five times a day forever and smack your heads on the ground and NO WOMEN 'cause they can be distracting sorry about the all-caps. And you, over there, I never ever ever want to see the bottoms of your feet, because have you seen what's all over the ground? Gross! And you other guys never cut your hair, and you other other guys cut all your hair off. And ladies, frankly, you are totally distracting all the time, what was I thinking? and you folks, if I catch you celebrating birthdays or anything else, for that matter, it's all over for you, and you other folks, happy happy happy and no medicine or doctors or pain meds or stitches or nothing), and on and on and on ...

Seriously? That's what God does? No wonder science thinks religious folk are idiots. They kinda are.

But it's kinda science's fault. Because science told us that in order to be smart, we had to have rules.

So we still think it's all about the rules.

OK, that's not really fair for science.

Humans seem to have this compulsion to come up with rules. Always have.

But it's really about the Interaction.

The rules come from the interaction.

Because the interaction is about relationships.

And good relationships need good rules.

Just like a working universe needs good rules.

Not to restrict the universe from doing things.

But to cause, to allow, to enable the universe to interact well.

So if there was going to be a real revelation from the real God, then that revelation was really going to be about relationships.

Because that's how we interact.

First, an interaction between God and humans.

And second, an interaction between humans and, uh, other humans.

Now watch this. It's going to get dangerous.

It's also going to be some math.

Interaction = relationship. That's first. That's from science.

Second.

The highest form of relationship = love.

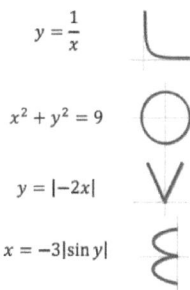

I just made that up, but I think it works.

Interaction, as I ponder it, has two forms.

Attractive and repulsive.

Gravity is an attractive interaction, where by the warping of space-time, objects in a gravitational field (which has infinite reach, btw) seem to be attracted to the center of the field, the center of the massive object. Dark Energy seems to be repulsive - massive objects are repelled from each other. Dark Matter, like regular matter, is attractive via gravity. The Strong Force creates a field via the gluon that attracts quarks together, and protons and neutrons together. Electro-magnetism is both attractive and repulsive. The Weak Force binds subatomic particles together into more complex elements under tremendous heat and pressure (which sounds like love to me) and has something to do with beta-decay, whatever that is, but it sounds like a broken heart, frankly. JK.

Quantum mechanics by its very nature is interaction between particles, communicating, changing each other in instantaneous and seemingly magical but ultimately entirely natural ways. But because it's Quantum and has to make life difficult, it's neither attractive nor repulsive. It is, however, purely interactive. Again, the poetry of QM is like a love song.

That might be a stretch. But really, the quantum world needs us as observers, and we need it because we are made of it. It is as interactive as interactive gets in a seriously interactive sort of way.

Human relationships are attractive and repulsive. We like each other, we dislike each other. We are attracted to or repelled by each other.

There is love, and there is hate.

So I will arbitrarily and randomly assigned values to those. Hate is generally the worst.

Love is usually the best.

And love, at its purest, is sacrificial.

Parent for child. Lover for beloved. Patriots for country. Superhero for humanity and the earth. Lassie for Timmy.

We are told that there is no greater love than for one human to give his or her life for another.

If you have ever truly loved, then you know the rich profundity of sacrificial love.

And thus, the interactional God who created the universe, who for some bizarre and inexplicable reason seeks an interaction with his creation, with his pathetic little otherwise pointless humans, and wants somehow to interact with us in a way that feels very much like love.

So third, highest form of love = sacrifice of life.

Now, go find a revelation that looks like that.

Chapter 32 There's Something There – Part 6

Evidence plus experience plus faith plus revelation equals something that might be a lot like God.

We talked about evidence. We talked about revelation.

Experience.

Oh, btw, I should mention in passing that *you* get to decide which way the evidence points (as we have said) and *you* get to decide whether or not any particular revelation is in fact a revelation, and which one you like and which ones you don't like.

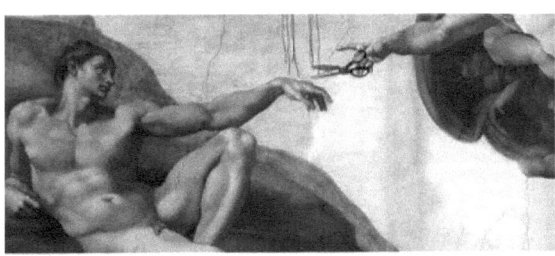

That seems to be built into the system. You get to decide. Now, whether or not that's a Free Will thing is something else you get to decide, unless you don't believe in Free Will, in which case, well, you don't have a choice about anything. Not only is the universe and everything in it an elaborate illusion, so is Free Will.

What THAT means is that you probably shouldn't feel too cocky about what you believe. Don't get the Big Head. If there's no Free Will, then really, getting all impressed with yourself is just silly.

Of course, if there is Free Will, then chances are good that's evidence that the Big Guy really exists.

In which case, getting all impressed with yourself is also just silly, since the Big Guy is, um, a lot Bigger a Guy than you are. Just sayin'.

It's also possible that the Big Guy exists even if Free Will does not.

Of course, if Free Will does not exist, then neither does love, and we're kinda bettin' on love.

(I should say that this week, science doesn't believe in Free Will. But it kinda goes back and forth on that.)(Evidence keeps changing.)(Darn that evidence.)

And btw, if you perchance don't believe in free will or gods or God, and you think religious people are idiots, well, frankly,

THAT makes no sense at all. More on that later.

And now we're back to talking about experience.

Let's go back to revelation for a minute.

There are lots of potential revelations in history to consider. Every religion seems to have one or two or several.

Islam's got one. Judaism's got a baker's dozen (I didn't actually count, so we're just approximating on that one.) Christianity has all of those plus, you know, the whole Jesus thing. The Latter Day Saints got one, in addition to whatever they pick and choose from other places. JWs got some, but they mostly overlap with Judaism and Christianity. Nearly everyone with a substance abuse issue has got a few. All the cults do. Without a revelation, well, what a crappy cult.

So how are you supposed to figure out 1) which one is the real one? and/or 2) are ANY of them a real one?

Yeah. That's a good question.

I have a close friend named Josie who if she's reading this right now is going WTH?! but I won't use her last name, so only I and she and all of our mutual friends will know. (BTW, in WTH, the H is Heck.)(It's a concession.) (Remember, "crapload" is our official limit.) (If you want to  put a different, alternative letter of your own creative choosing in there, well, if there's Free Will, go right ahead, and if there's not, then you can't help it anyway, so go right ahead.)

Anyway. Short version. In high school (a very long time ago) she fell in love with another friend named Jeff (who's right now going WTH?! Sorry, dude. It was some significant fraction of a century ago, and it was kinda sweet.)

Josie was, and remains as far as I know, an atheist.

But she told me at the time that the love she felt for Jeff caused her to say, this is so powerful and incredible and uncontrollable that she could understand how God might exist.

I think she got over it. Sorry, Jeff.

But 1) since the universe is only and always formed of interactions and 2) the highest form of interaction is (we are saying) love, then 3) the experience of love might be that experience, the revelation that aims us in the direction of God.

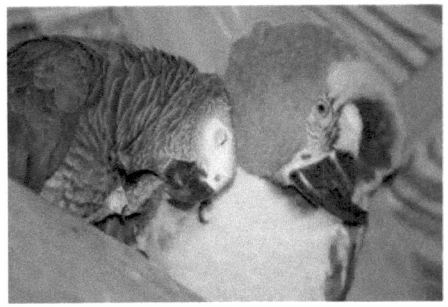

Now love might just be biochemical and a evolutionary response to the need to procreate and keep the species going and Free Will might not exist and it's not really love it's just urges, but

I don't think so.

Those things might be true, but to say that's all that love is ... reductive.

It might be instinctive (whatever THAT means) and evolutionarily derivative and all of that, but

it's more.

You know it and I know it.

Yes, love is biochemical. So what? So are we all.

But Complexity Theory tells us that there's a tipping point (derived from Chaos Theory) wherein something ...

happens ...

And something magical emerges from the mix that is far more than just biochemistry.

Yes, procreation, yes, keeping the species going, sure.

But love is vastly more than just biochemistry or an evolutionary imperative.

Love is profoundly, magically, mysteriously, wondrously built into the fabric of existence, the ultimate form of interaction in a universe that is defined from smallest to largest, from the least to the greatest, from the most transient to the most persistent, from the quantum smallness to the relativistic vastness of time and space, from the relationship between particle and observer to the relationship between space-time and matter by

Interaction.

Ultimately, interaction is experience. Everything that interacts enters into an experience with the focus of the interaction, and reality emerges and is defined by that experience. It is Quantum and Relativistic and Newtonian. It is transcendent.

Wow. That was a bit much. Let's rephrase that.

It takes two to tango.

That is, it is the interaction between two dancers that creates the dance. Without the interaction, no tango. Without two dancers, no tango. Without the tango, no dancing, no dancers.

And no experience.

If God does not exist, then there's no interaction, no experience. Just delusion.

But if God exists, and if he is interacting with the universe, and if he is a God of love, then it is possible to interact with that God and experience love.

It may inevitable.

And it will be different for everyone, unique to each person, but somehow universal all at the same time.

Unique. And universal.

That's what love is.

And the love that we all feel for, I don't know, cat videos on YouTube, or, more seriously, the people whom we each love, just might be that space-time-centric, Newtonian, Quantum, Relativistic, Chaotic and Complex interaction that points us to, I don't know,

the actual real moment when the 5th dimension (or whatever) touched the 4 dimensions that we inhabit and divine love became tangible but that would take some faith. So maybe we should talk about it.

Chapter 33   There's Something There – Part 7

There was this very nice British teacher at an international school I worked at once near Geneva where he taught a class called Theory of Knowledge (TOK) - I spend most of my lecturing time in TOK settings all over the planet - and he was talking about faith one day.

He said that faith was like this: you could believe in little green men flying around the room, or anything at all, really, and that would take faith, but clearly zero intelligence, so faith was believing in ridiculous things and was clearly for stupid people.

In the nicest possible way.

And, he said, since you can't offer him any tangible evidence that there in fact might be little green men flying around the room (apparently desperately looking for little green flying women)(I added that part just now.), then he is under no obligation or compulsion to believe anything except that you are a raving lunatic.

And so, faith is ridiculous. And for stupid and/or crazy people. Faith is just a short hop from insanity. I added that last part, too. But it's true. And thus religion is ridiculous, since there's nothing that is actually true about it, and if you are religious, then you are ridiculous.

In the nicest possible way. Because he was British and very nice about it all. Then we had tea.

Actually, because he was very nice, he then let me talk about faith to his 3 classes. And, because he was very nice and kinda impressed, he invited

me back thrice annually for the next thrice years. The Brits say "thrice" a lot. I don't know why.

Here's what I said.

First, believing in little green men, or in anything, would be crazy without at least some evidence. Something that is true. A fact or two. Little green droppings scattered around the room, for example.

I'm not even going to mention that a lot of very smart, generally skeptical-about-religion-and-a-lot-of-them-actual-atheists scientists believe wholeheartedly in the Multiverse and/or String Theory and/or

Loop Quantum Gravity and that the universe might be a hologram and/or a computer simulation and that there really really just have to be aliens out there somewhere, and a bunch of other sciency sounding things ...

Without any evidence whatsoever. No evidence. Not a single dingle fact. Little green men. That's what we're talking about. Little green men. And women, too, I suppose. Without any facts, you can believe in whatever you want. Right?

Anyway. Sorry for the tangent. Pay no attention to the tangent.

So. If you're gonna have faith in something, there needs to be a fact or two.

And so, there are some actual facts. Buddha lived, was a real person. Confucius, too. Mohammed. And his wife. Wives. Moses, real guy. David. And his wife. Wives. Jesus. Peter, Paul and Mary. The Bible ones, not the folk singers. OK, the folk singers were real, too.

All of these folks lived, did things, had lives, left records (with apologies to Peter, Paul and Mary), are part of history.

So for example, let's take the one that is common to many faiths, including Islam, Judaism, and Christianity. And Hinduism. LDS. JWs. Not Scientology. That's whackiness without facts, little-green-men-ism, bonkers, nutso. Christian Scientism, though. Some others.

So for example, let's, in what is an obvious attempt by me to take the conversation in the way I'd like it to go, take Isa. Yeshua. Joshua. Jesus. A rose by any other name yadda yadda yadda.

So here's the thing. Isa/Yeshua/Jesus was a real guy.

Born. Raised. Lived. Said stuff. Did stuff. Had a big impact. Had to have, since Islam, Judaism, and Christianity (and its denominations and occasionally weird off-shoots and strange detours and often cultic bizarrenesses and not nearly rare enough white supremicists) all have had to deal with him, each in their own way, each different from the others. But believe him or not, none are saying he was just a myth.

That's just history. That's not religion. He was a real guy with a real history. You don't need faith for that. That's just a fact.

He also died. Everybody did, so that's not unusual.

Before that, history tells us that he was arrested as a rabble-rouser (governments hate rabble-rousers), tried, convicted, executed, entombed.

That's history.

And his body was never found.

That's history.

What happened to it, is where faith begins. Where religion starts.

So faith is about facts, about evidence, but necessarily incomplete evidence. And the faith parts tend to be what science would call "supernatural", that is, not part of nature, outside of the laws of science.

Now there's evidence about what happened after the body disappeared, but it's evidence just from Christians. So it's biased. It was a supernatural event for which Christians claim to have real evidence from real people.

And you get to believe it, or not. That's faith.

Do you get the whole story? Hardly ever.

But that's also true in science.

For example. How many stars do you suppose are in our galaxy?

Maybe you've been told. It's somewhere between one hundred and three hundred billion. Or 400 hundred. Hard to tell.

How many stars can you actually you yourself with your eyes see at night, if you could count all the ones that are visible by the naked eye from earth on a really dark night?

Here's the number. 9110. Nine thousand one hundred and ten stars.

Not 9111. Just 9110.

In 1888, we thought there were just 6188 stars. So we're getting closer.

Now I've never counted stars at all, so I'm taking both numbers on faith in the people who tell me things about stars. The funny thing is, we don't really know how many stars are in the Milky Way, but we do know how many we can actually see at night. 9110. We know there are craploads more stars than that in the Milky Way, but not how many more.

And that's just our galaxy. Wanna know how many galaxies there are?

Somewhere between 100 billion and a trillion or two. Or twenty. So that's plus or minus 900 billion. Or 19+ trillion.

Wanna know how much bigger the actual universe is than the part that we can see with our actual eyes? OK, that's not right. We can only see 9110 stars with our actual eyes, and no galaxies, not really.

So we can't even see the part of the universe that we think we know about, much less the part that we will never be able to see and will never have any idea how big it is.

Scientists tell us that their experiments tell them that we'll never know. And they tell us things that their experiments tell them we do know.

And we take it on faith.

And then when they discover that, oops, they didn't quite get that right, but this time we've done a much better job, we take that on faith.

Lots of it doesn't make any sense really to us or to them. Space-time bending and warping. Particles everywhere all at once. Black Holes. Big Bang. Quantum Entanglement. Quantum Tunneling. Quantum Everything. None of it makes sense, it all seems quite miraculous, the math works, the experiments work, but it's all so so so so weird!

But they have faith in the math and the experiments. Ultimately, they have faith that the universe makes sense, that they can figure out a lot of things, that there is order and structure and math that describes it all.

They even talk about math being a kind of miracle. Feynman said that.

So the universe as it actually is, is weird, bizarre, counter-intuitive, almost nonsensical. But our evidence plus our experience plus our faith tells us that it's ok to believe in science and nature and the universe, even though it's not really here at all, except for maybe ...

Interactions.

Now, where scientists and skeptics and atheists take religious folk to task is over things like miracles and supernatural stuff. And honestly, religious folk are prone to seeing miracles in every parking space. Every time something good happens, whee! It's a miracle!

'Course, for the non-believers, every time something bad happens, it's God-is-a-total-jerk time. Or he would be, if he existed.

So we should talk about that.

Chapter 34   Gaps

I find this to be a curious thing.

Atheists think that religious people are idiots for believing in miracles. Atheists also tend to be mad at the God they don't believe in for allowing evil things to happen.

That is, for not doing miracles. To stop the evil things from happening. Of course, if he did stop some, they'd never know.

And, as it turns out, they are desperate for God to stop things like tsunamis or genocides or earthquakes or tornadoes or serial killers or the Kardashians, but not the evil things that they themselves do.

And then religious people, for whom finding a good parking spot is a miraculous Act of God, somehow don't notice when God doesn't do things like stop tsunamis or genocides or earthquakes or tornadoes or serial killers or the Kardashians.

We are a confusing species. So let's try to figure it out.

There are always things humans do not understand.

As religious people (and humans have always been religious), when we met things we did not understand, we would ascribe those things to God, or the gods.

Volcanoes, earthquakes, sun, moon, stars, cats, mysteries of all shapes and sizes.

Any gap we found in our ability to explain the universe, we filled with gods.

Later on, science would call this the "God of the gaps" and use it to dismiss religion entirely. This they did from the entirely reasonable perspective that science had explained most of the things we didn't understand and had blamed on God, or the gods. And they felt pretty good about being able to explain all the things they hadn't been able to explain yet. They had a good track record.

They have a point. And we haven't learned the lesson yet.

It's pretty simple. Bad things would happen. You know, volcanoes, earthquakes, tsunamis, floods, famines, plagues, fires, landslides, cats. Terrible things. The Kardashians. Teeeerrible things.

Since it didn't make sense to us that bad things just happen sometimes, we would figure that the gods were angry about something we did and were doing these bad things to us as punishment.

Since we didn't know what we had done, and the gods weren't letting us know, weren't giving us any hints, we would make stuff up.

And then we'd try to fix it by doing, I don't know, whatever. Sacrifice some virgins. Cut out some hearts with black obsidian knives. Shrink some heads. Eat some enemies. Whatever seemed right at the time.

The gods were fairly irrational about doing bad things to us, so we were pretty irrational about trying to keep them happy and off our case.

There was this sense that we had to sacrifice stuff to the gods to keep them off our backs. Gods - very demanding, very much like little babies, teenagers, bad bosses, mother-in-laws, and Kardashians. It's always totally about them.

The Greeks spent a lot of time and energy trying to keep the gods happy. The Romans, with pretty much the same gods, did the same thing.

Even in our more modern, educated, sophisticated times, religious people do the same thing.

Hindus have some 350 million gods they have to keep happy by following their dharma to build up good karma so that reincarnation can happen at a higher level next time around and keep them from harma. Ha. Sorry.

Buddhists, at least in its purer, Buddhist form, have to avoid developing fond attachments for the things of the world, including friendships and family relationships. In its more western, less Buddhist form, there are chantings and incenses and good feelings and thoughts.

Jews have something over 600 laws they have to follow to keep God happy. They have each their rabbinical interpretations of how much to carry and how far to walk and who gets to wear what when and why. It's not really a growth religion because, well, circumcision.

Muslims, luckier than the Jews, have about five laws, but there are myriad interpretations of how to follow those laws, and then they've got others.

Outside of the mainstream religions and denominations, cults are defined by the weird, cultish things they demand of their followers to keep the gods happy.

And Christians spend a lot of time and energy trying to keep God happy, trying, as it were, to stay saved.

They mostly try to do this like everyone else does it - by trying to follow laws and rules and regulations, even though they know better.

Because Christians and Jews and Muslims and Buddhists and Confucians and Shintos and Jains and Hindus and everybody in every faith and religion and belief system EVER thinks that what God is all about is

1) being pissed off all the time because apparently none of you IDIOTS can

2) follow all the rules.

God doesn't need to use all-caps, btw, but, well, you know.

And atheists and scientists very rightly think to themselves, God is an idiot, because all he does is demand that people follow of bunch of mostly stupid rules. And he's mean and nasty, besides. If he exists at all, that is.

So the atheists and scientists reject the existence of *this* God. This mean, nasty, petty, tyrannical, cruel, heartless, arbitrary God who clearly and easily could have created a universe that didn't have evil and suffering in it and should fix it anyway, although that would take miracles, which they don't believe in, because those would be supernatural, and they don't believe in supernatural stuff. Except for Big Bang and Quantum Theory. Hmm.

Now. Just suppose. Maybe that's the wrong God entirely to believe in, or to reject. Maybe most of us, believers and non- alike, have God got mostly wrong.

Maybe he is not a God of rules and regulations. Maybe he is a God of interactions and relationships. Maybe he is a God of love.

Humans get it backwards. We traditionally have thought that if we follow all the rules and regulations, then God will love us and will not squash us like a bug. If you want an equation, it might look like this:

Rules => Love.

Or like this: Obedience => Love.

And like this: Disobedience => Getting Squashed Like a Bug.

Instead, maybe it looks like this: Love => Interaction => Rules.

So instead of Acting Right so That God Will Love You, maybe it's

God loves you already, and wants you to love each other, and here's how you might do that. You interact with each other, and out of the interactions come the standards of behavior.

Culture emerges from relationship. The interaction between me and thee.

Just like, gravity emerges from relationship. The interaction between space-time and matter. Matter emerges as an interaction, energy becoming matter becoming energy becoming matter. And so on.

The universe arrived with four, maybe five rules in place, and the rules are all about interaction. Gravity. Strong force. Weak force. Electromagnetism. Quantum Mechanics, whatever the heck that is.

QM might just be the miracle juice we need to make everything happen.

And all the rest of the laws of nature emerge from that starting point as a direct by-product of interactions.

Four rules. Four types of interactions. And the supernatural, the magic, the mystery, the wonder of the quantum universe that gives us something from nothing and a universe that is nothing but connected interactivity between forces and particles. Here, read this:

Reality is Relational by Carlo Rovelli, *Reality is not what it seems* (2017)

"The third discovery ... is the most profound and difficult...

"The theory does not describe things as they "are": it describes how things "occur," and how they "interact with each other". It doesn't describe *where* there is a particle, but how the particle *shows itself to others*. The world of existent things is reduced to a realm of possible interactions. Reality is reduced to interaction. Reality is reduced to relation.

"In a certain sense, this is just an extension of relativity...Quantum mechanics extends this relativity is a radical way: *all* variable aspects of an object exist only in relation to other objects. It is only in interactions that nature draws the world.

"... there is no reality except in the *relations* between physical systems. It isn't *things* that enter into *relations*, but rather *relations* that ground to the notion of *thing*.

"The world of Quantum Mechanics is not a world of objects; it is a world of events. Things are built by the happening of elementary events.

"A stone is a vibration of quanta...just as a (ocean) wave maintains its identity for awhile before melting again into the sea."

What the universe does emerges from the interactions. The culture of the universe emerges from the interactions.

How and why what it does, and how and why what we do as humans, as life-forms, as the aliens' aliens, emerges from our interactions.

If we start our interactions with an intention to have that interaction be sort of, you know, nice, um, loving and kind and good and thoughtful and sacrificial and caring and all of that, then what emerges is highly likely, statistically, probabilistically, scientifically, experimentally verifiably, to be, you know, kinda

good.

On average. Over time. Generally. Usually. Interactions are complicated things, which makes it all just that much more interesting and sometimes messy and fascinatingly unpredictable and all butterfly-effecty and emergent and stuff.

But still. Worth a shot.

Chapter 35   Gaps Are Back

Now, what happens to science when the gaps come back?

Remember the God of the Gaps problem?

Religion said, we got earthquakes and volcanoes and plagues and floods and fires and all sorts of horrible things plus Kardashians because the gods are all pissed off at us. Also, rainbows and butterflies and waterfalls and suns and moons and stars and sunsets all come from the gods. So throw some virgins into the volcanoes and we'll get rainbows and sunsets all over again. The gods love it when we throw virgins into volcanoes. Those crazy gods.

So science, all Newtonian about it, says, hah! you guys are idiots. Everything happens because the laws of physics make everything happen. Gods!?!? Seriously!?!? You guys are idiots.

And we were, kinda.

'Cause the laws of physics DO cause everything to happen.

If we had not been idiots, we might have said, um, so, where do the laws of physics come from?

And THEY would said, hah! you guys are idiots. The laws of physics have always been here. Just like the universe has always been here. ("Oops" is what they said about that later.)

How many laws of physics are there, actually? we could have politely asked.

One, they would have said. Gravity. Everything else comes from that.

Then after awhile, they would have said, Two. Gravity and Electromagnetism. Everything else comes from those.

163

And then the 20th century rolled into view, and all of sudden, things got messy.

First, there was Quantum Theory.

Then, there was Relativity.

Then, there was Big Bang.

Then, the universe hadn't always been there.

Then, the laws of physics hadn't always been there, either. Because nothing had always been there.

Plus, then there were two more laws. The Strong and Weak Interactions. That's what they call them.

And all of a sudden, there were more Gaps that needed to be filled with science.

And the actual science that filled the Gaps just gave us even more Gaps. Gaps upon Gaps.

Fortunately for science, religious people didn't like Big Bang, so it turned out not to be much of a problem, because then they didn't know about the new Gaps.

But it should have been.

Because science had filled all of the OLD gaps with Newtonian science, which doesn't work at all well at filling the NEW gaps.

And we have all of the NEW gaps because of the NEW science. Relativistic, quantum science. Which created the gaps.

It's a problem. Because the new gaps look like this:

Where did all the science come from? Laws of physics-wise, I mean.

Where did the universe come from? Since there were no laws of physics to make it come from anywhere, and since there was no where there for it to have come here from there from.

And what the heck is this whole quantum thing, anyway?

So now we've got quantum weirdness and space-time weirdness and Dark Energy and Dark Matter and Strings and Loops and Holographic and/or Computer Simulated Universe options and Biocentric Universe options and 41 (or more) interacting universes options and Multiverses options and Higgses and Cosmological Constants and Inflationary options and there's hardly anything we know about a lot of them that might be true and a lot that we know that is true that is totally bizarre and NOT Newtonian.

And frankly, there's a lot more gaps than not gaps these days, cosmology-wise.

And God has become a reasonable answer, unless

you think that God will never ever be any kind of reasonable answer because

Gaps.

But, here's a thought.

Just because God was a lame answer for some problems we had in understanding nature doesn't mean

That he might not be a really fine answer for other problems we have in understanding nature. Maybe we just need to ask better questions. And maybe we need to be ready for our understanding of God to be too small.

Which only makes sense, since our understanding of the universe has clearly been too small.

'Cause here again is what science did. It assumed that the God it *didn't* believe in was exactly like the God that religious people *did* believe in. That since we knew that volcanoes weren't actually the gods blowing off angry steam, that there was a nice neat scientific explanation for volcanoes, then THAT God or THOSE gods did not exist.

And indeed, they did not.

But a much better conclusion would have been,

Maybe God is a bit more complicated, complex, and interesting than the pissed off gods of yore. And maybe rejecting the existence of God today by assuming that God is the same God or gods that the modern religions accept today is no more insightful or imaginative.

That is, to say that the God of the Old Testament or the Torah or the Koran is harsh and brutal and therefore God does not exist, is the province of small minds.

And maybe assuming that God is constantly in a state of being constantly pissed off is a bad assumption.

The volcano virgins of the world like where this is going.

Chapter 36   Free Will or Free Won't

Let's us talk about free will for just a bit. And we'll start off being a little irritating and obnoxious, as is our wont.

1, 2, 3 go:

Your opinion about free will is entirely irrelevant.

Sound familiar?

Free will either exists, or it does not.

What you *believe* about free will is immaterial.

Your belief in free will does not cause it to exist, nor does your disbelief cause it not to exist.

Nor is pretty much anything that science has to say about it very informative.

I posted two articles in the same week once on FB, one of which said free will exists, one of which said that it did not.

There will be a study that suggests that it does not, and some time later, another study poking holes in the first study. I've been watching this thing for 25+ years, so trust me on this.

That's the way science works, of course, and that's a good thing.

But the way the human mind works is to pay very close attention to the study that we like, and not so much to the one we don't like.

Plus, the way things get reported is that first study gets large print on the front page (FREE WILL DOESN'T EXIST! PROOF AT LAST!) and the second study somewhere else in tiny tiny print (Free Will Might Exist. We were wrong. This time. But you just wait. One of these days...).

The problem with figuring out free will is that it's not just or even mainly a philosophical or theological question,

It's a brain question.

And here's what we do not understand at all.

The brain.

Here's a bit from an article in Wired Magazine, the world's greatest magazine ever:

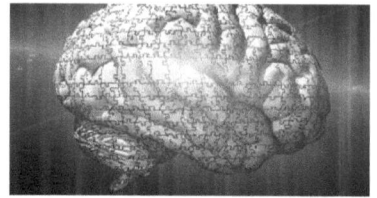

"Studying the brain now is like trying to navigate a vast city without any driving instructions. You don't know where you are, and you have no idea how to find what you're looking for.

"Every brain is profoundly unique, a landscape of cells that has never existed before and never will again.

"The same gene that will be highly expressed in some subjects will be completely absent in others.

"This variation is even visible at a gross anatomical level - different people have differently shaped cortices, with different boundaries between anatomical regions.

"If the human atlas is like Google Maps, then every mind is its own city.

"Scientists assumed for decades that most cortical circuits were

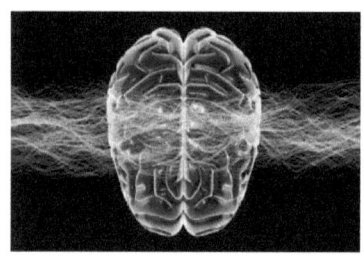

essentially the same - the brain was supposed to rely on a standard set of microchips, like a typical supercomputer.

"But the atlas has revealed a startling genetic diversity; different slabs of cortex are defined by entirely different sets of genes.

"The supercomputer analogy needs to be permanently retired.

"Scientists are just starting to grapple with the seemingly infinite regress of the brain, in which every new level of detail reveals yet another level.

"The problem with this data is that it's like grinding up the paint on a Monet canvas and then thinking you understand the painting.

"You can't help but be intimidated by the complexity of it all. Just when you think you're getting a handle on it, you realize that you haven't even scratched the surface.

"What you mostly discover is that the mind remains an immense mystery. We don't even know what we don't know."

So when some guy in a lab is trying to figure out whether or not there is free will, he has to test living human brains in living human heads, brains that he might even assume that he understands,

But he doesn't. Or she. They. All of them.

And when you or he or she or they or all them read some article about some other scientist(s) claiming they have figured out something about the way the brain works, the truth is, they've made assumptions about the way the brain works that don't seem to be at all, um, assumable.

So if our first guy or whoever is trying to figure out if there's free will, he/she/it/or they is doing a free will test on something (a brain) he/she/it or they have no understanding of.

So really. It's just not going to work.

The interesting thing is how badly we seem to want free will not to exist.

Someone's gonna have to explain that one to me.

Oh, never mind. I'll just do it myself.

Chapter 37   A Brief History of Free Will

Rather than doing actual research, I'm just going to make this up.

If free will exists, then I get to do that. Make this up, that is.

If free will does not exist, then I apparently have no choice but to just make it up.

So. It always felt a lot like we had free will when we still thought we had free will. I mean, there was no question.

Were making it up. The world, the universe, life, reality. Especially reality.

– Tom Robbins –
OkyDay.com

And punctuation mistakes. Making that up, too.

Right or left. Up or down. Do a bad thing, don't do a bad thing. Or a good thing. Eat this, don't eat that. Go here, go there.

My brain (speaking metaphorically about all brains) felt very much as though it was faced with decisions all the time and got to make up its own mind, since it was, in fact, a mind.

But then Newton came along and discovered that the universe worked according to laws and rules and regulations. All the time. Never stopped. Never took time off for a vacay. No coffee breaks or weekends at the beach.

And this meant that everything worked according to the laws. All the time.

And the laws had always been there. Since the universe was infinitely old and large and all of that.

You and me and everything and everyone are made of tiny little particles that work according to the laws. All the time.

Pretty soon, some smart guys said to themselves, huh. THAT'S interesting.

Because what that means is that since our brains are made of little particles that work according to the rules all the time, then ...

What we THINK is a free will decision is actually just something that the particles did. The laws had been working forever, making all the particles in the universe do stuff, and eventually, some of the particles ended up in my brain making me say something stupid to this really hot girl in high school instead of something smooth and smart and urbane and debonair which would have in turn made all of her brain particles want to run off with me and pursue connubial bliss and live happily ever after.

So. All caps coming.

IT'S NOT MY FAULT!

Good to know. Sort of.

To review. Since the universe has always been there and the laws have always been there, then there's no free will. Ever. Never. Not possible. Not gonna happen.

And oh btw, we also don't need God to explain how things work, since the laws make everything work and the laws had always been there so we didn't need God to start the game, get the ball rolling, blow the whistle, wave the flag. The game was eternal, had always been the game.

So the smart guys said to themselves, huh. Sweet. No God. So we can do whatever we want. And no free will. So it's not our fault. The particles made me do it. Niiiiice.

There was one tiny little teensy weensy hardly-worth-mentioning problem.

Ahem. The universe had not always been there. Neither had the laws. Everything had a starting point.

Dang.

Here's where we are, then.

We THOUGHT we didn't have free will because the universe was infinitely large and old and the laws had always been there making us do things that we THOUGHT was free will but wasn't and history was an infinitely long string of laws making particles do stuff that included all the stuff that we were each doing that we THOUGHT was stuff we decided to do but didn't really.

But this was entirely totally absolutely dependent on the universe and the laws being infinitely old.

But. They aren't.

Which means that the universe and the laws used to not be here. At all.

So since we used the infinite universe to get rid of both God and free will, then since the universe isn't infinite, then the possibility that both God and free will exist, um, exists. It's possible. Not definite. But definitely possible.

You probably didn't know that's how we got rid of both God and free will. 'Tis, though. We sorta forgot that.

Then when Big Bang showed up (thanks, Albert. And Georges. You need to say *Georges* the French way. It's pronounced *baguette*. I may have mentioned that already, but I'm pretty sure you forgot.), all of sudden things got complicated. Science didn't even like Big Bang (Albert didn't even like Big Bang) because it meant that we would have to start talking about God and religion and going to church or temple or mosque or Mecca or Jerusalem or Rome or Tokyo (Tokyo?) and cleaning up our acts naughtiness-wise and

nobody really wanted to do that.

But it all worked out because religious people didn't like Big Bang either because (science was right) they're idiots. With apologies to religious people. OK, not really.

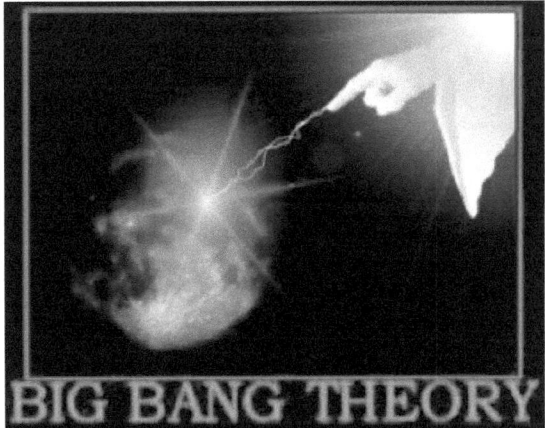

In fairness to religious people, we should probably mention why they didn't like Big Bang. They believe that God instantaneously created the universe out of nothing, but Big Bang Theory says that the universe suddenly popped into being out of nothing.

You can see the problem.

Yeah, me neither. I just don't get it.

Anyway. So. Free will might be back. That's not the movie about the whale, btw. Which, btw, my British friends think the title of which is hysterical.

To sum up. Free will (and God) went away because of the infinite universe, and they came back maybe because it's not infinite. Time-wise, that is. Space-wise, we'll never know.

And I'm out of time. But not space. Ha. Hysterical.

Chapter 38   Free Will is Back. Sort of.

So. Do we have free will, and does God exist?

Well, if you're going to get rid of them again, you're gonna have to have a darn good reason. Better than, I DON'T THINK SO! That's so lame.

Let's talk about it for a minute.

If you are someone who doesn't believe in free will (and feel free, btw), then you could be accused of not thinking clearly about it.

Because if you are also (as there is an excellent chance) a person who detests religion and all this God talk, then you are not making any sense.

Because (you know where this is going, don't you?) if there's no free will, then all of the religious people have no choice but to be religious. All caps warning. It's not their fault. I changed my mind about the all caps. You gotta lay off. Stop hassling them. The particles made them religious, ultimately.

And you, btw, are irreligious, atheist, agnostic, and/or skeptical not because you are so smart. You have no choice. The particles made you that way. Lucky you.

None of us get to claim any credit or take any blame for our looks, brains, success, talent, skills, experience, hard work, money, possessions, achievements, or anything. For our failures, shortcomings, flaws, zits, flatulence, body odor, obesity, baldness, bad posture, terrible grades, lousiness at sports, friendlessness, ugliness, or anything.

So if you or anyone are taking any credit or blame for anything and if you don't believe in free will, then you are no less an idiot than religious people.

Actually, to be totally precise, none of you are idiots. If free will does not exist. You only have the capacity to be an idiot if free will exists.

So if you think religious people are idiots, or if you think atheists are idiots, or if you think anybody anywhere for any reason at any time is an idiot, then, really, the only way for that to be true is if

(you know where this is going, don't you?)

Free will exists. And by extension, God might also exist.

WTH?, you might exclaim with some heat and passion. H, to remind you, is Heck, because crapload is our swearing limit. It's like a law of physics.

Because, we reiterate, the smart guys used the deterministic infinite universe to get rid of both Free Will and God. And the most recent group of smart guys have not been able (or motivated) to update that. Atheists are Newtonians all. I'm so embarrassed for them. You. Whatever. You're still using a Newtonian universe to get rid of God, unbeknownst to you. You also think that God doesn't exist because religious people are idiots, which, I remind you, is only true if Free Will exists, and if

Newton. Not Newman. Pay attention.

Free Will exists, then so does God. It's like Catch-22, which I should not admit is my favorite book of all time.

Major Major Major Major. I just had to say that.

You should be nicer to idiots, btw. If free will doesn't exist.

If it does, then, well, they're idiots. Still. Better to be nice. There's that naughtiness thing lurking out there somewhere. Plus, we are each an idiot about something. Einstein said that. Paraphrased.

Symmetry. By Da Vinci as a child.

Summing up. The only way that religious people are idiots is if free will exists, and if free will exists, then (it is highly likely) God exists.

So summing up again. The only way that religious people are idiots is if God exists.

That's some nice symmetry.

So if free will does not exist, then you can't be an idiot (nobody can) and nothing is your fault and making fun of idiots or someone else's beliefs is not a fully informed thing to do. Though, of course, if there is no free will, then making fun of others is not your fault.

Yeah. See. Nobody lives like that.

We completely and totally act like people actually are idiots and things are someone's fault, otherwise nobody could ever go to jail for doing anything wrong or evil, and in fact, evil would not exist. Good, neither. There would be no good or evil, because for you to do something evil or something good would take a free will decision on your part, which you could not do if free will doesn't exist. The particles made you do it.

So we all spend all of our time acting like we all have free will. Even the scientists who think we don't have free will, act like they have free will.

Weird.

That doesn't mean we have free will, though.

177

There's this Newtonian concept called Determinism, and another one called Mechanism. They are the problem.

Determinism says that everything is predetermined by things that happened already. Again, it's all about the particles.

Mechanism says that, because of the laws of physics, everything acts like a machine, and what a machine is, is predictable. Everything happens the same way that it has always happened, so we can predict what's going to happen.

So the laws tell the particles what to do, whether they are single particles or a galaxy made of particles, and the particles do just that.

And since you are made of particles, theoretically we can look at the way you have ever done anything and predict what you will do next.

There you go. No free will. Even in a universe with a starting point.

Since we still have to answer the question of where the laws and the universe came from, God is still theoretically around even if free will is not.

But.

This is all Newtonian. That's not the universe we live in. Well, it is, in part, but only the dull and boring and predictable parts.

It's really a Quantum thing.

It's pretty simple.

The Newtonian thing says, if you know everything about the particles (where they are, how fast they're going, stuff like that), then you can predict where they will go and what happens when they get there.

And you lose your free will. Because you are made of particles.

But the Quantum thing says (wait for it),

you cannot know everything about the particles. Any particle. Not even one.

It's called Quantum Uncertainty. You can't know where they are, or if they are, or where they're going, or if they're going. Because actually, they don't have a where yet. It's not like they are someplace and you just don't know where it is. They aren't anywhere yet.

And since we can't predict what the particles are going to do, and you are made of particles, then theoretically we can't predict what you are going to do.

And now. Theoretically. You get your free will back.

We're not really sure what that means, of course. That's the problem.

Chapter 39 Free Will and Butterflies

The problem is, the universe is still pretty predictable. Like, clockwork.

And so are you.

We can predict lots of things the universe is going to do. Sunrises. Sunsets. Eclipses. Rainbows. Lots of things.

We can predict lots of things you're going to do. Given enough information. You'll eat. Sleep. Poop. Grow. Learn things.

But the other problem is, we can't predict everything. Most things. Lots of things. But not everything.

So we don't really know if you have a little bit of free will, or a lot, or almost none.

It's hard to tell.

There's this thing called Chaos Theory. It says that although the universe and nature and stuff are MOSTLY predictable, every now and then, things get really really unpredictable.

It's called the Butterfly Effect. It says that sometimes, something as small as the flapping of a butterfly wing can set off a series of events that cause something HUGE to happen. Like a hurricane or a tornado.

Every now and then.

So mostly you are predictable (you and the universe), but every now and then,

You are really really unpredictable.

Does that mean you have free will?

Ah. No. Not completely.

It says that just because we didn't know what made you do whatever weird, bizarre, possibly illegal and immoral thing that you did, there could have been something that made you do it.

So although it's possible for you to have free will, we still don't know if you actually have it.

What there is, is the possibility for free will to exist.

Because the universe is not completely deterministic, thanks to quantum things and chaotic things, it's possible for you to have free will.

It's clearly not predetermined that you have free will.

Ouch. That's confusing.

Because here's the thing. Most of the time, we act like we have free will, but whether or not we are actually making free will decisions is hard to say.

Mostly, we don't.

That is, we are little puppies slobbering when the bell rings.

You know. Pavlov's Dogs. All of that. That's when this guy named Pavlov (great name for a Star Trek character) trained a bunch of dogs to salivate when they heard a bell ring by ringing a bell and feeding them doggy vittles.

After awhile, all he had to do was ring the bell, and looky there! Doggy slobber. Even without vittles. After another while, eventually one would guess that the doggies would figure out that they weren't getting any snacks and stop slobbering.

We're like the dogs. We are (largely?)(totally?)(that's the problem - don't really know) influenced by cultural bells.

We are (largely?)(totally?) influenced by our culture. By our friends. Magazines. Movies. TV shows. Celebrities. Families.

We dress the way we dress, cut our hair, decorate our homes, buy the things we buy, read the things we read, listen to and watch what we listen to and watch because of the little bubbles that we live in.

Our hearts also beat and our lungs breathe and our entire bodies work constantly without us making any decisions about any of it.

That's a good thing. If we had to constantly think about keeping our hearts beating and our lungs breathing and our spleens spleening, we'd die at night.

A lot of the things we do when we're awake and asleep, we do on autopilot. We don't think much about it. We have trained ourselves to walk, talk, drive, run, jump, eat, and do all the things we do without thinking about them very much.

And that's a good thing.

But the question is, are we (largely?)(totally?) controlled by our autopilots? What are the decisions that we make? Are (none?)(some?)(any?)(all?) of them free will decisions?

Turns out, it's really hard to tell.

Behavioralist BF Skinner said that none of them are. He called it "conditioned responses". Everything we do, we do because we have been trained or conditioned to do so. Even, apparently, become famous behavioralists.

I think that's a bit of a stretch.

But it's really hard to tell where free will actually starts, since we are so heavily (completely?) influenced by outside factors, outside of our ability even to be aware that it's happening. We are truly conditioned to do most things.

The question is, are there things that we do outside of conditioning? As free will decisions?

And the answer is ...

You have the *potential* to do things that you are not conditioned to do.

Whether or not you actually ever do that, is the question.

And the answer is, maybe, every now and then. Like a hostile butterfly.

Maybe, every now and then, we do something unpredictable. We make a free will decision. Maybe.

Note: "According to "Ivan Pavlov: A Russian Life in Science," a biography by Daniel P. Todes, Pavlov never used a bell with his dogs…'(he most frequently employed a metronome, a harmonium, a buzzer, and electric shock)' … Pavlov's dogs weren't salivating at the sound of the bell, er, the buzzer at all; they had been conditioned to start drooling for food when they saw the white lab coats." Curiosity.com, The One Thing You Know About Pavlov and His Dogs is Wrong, April 6, 2018, Joanie Faletto

Chapter 40   Free Will and Harley Guys

It would not be wise to piss off Harley guys.

However. If any happen to be reading this, they'll be the ones with a sense of humor.

We're counting on that.

Harley guys ride Harley Davidson bikes. Harleys (as in, the bikes) are all pretty different. You can get them built to your own specs. So that, you know, it's not like all the other Harleys.

And then you get dressed to ride your Harley. Then you ride off to meet with all the other Harley guys.

Might be your own little group from the local Harley hood.

Might be Sturgis. Which is the annual mating grounds of Harley guys. Don't Google *Sturgis*. Disturbing pics.

Harley guys are fiercely individualistic. They set themselves apart from regular guys. They are fiercely different. They are not like anyone else. They stand alone. Fiercely.

Except.

They all kinda

look alike.

So do their bikes.

They fiercely want to look different from all the other guys.

But not that different from, well,

each other.

Because they want to belong to that group. They want to be a Harley guy. So they gotta ride a Harley and dress like a Harley guy.

Even gotta have a Harley chick on the back. And she's gotta dress (or not) like a Harley chick.

So here's a subculture that fiercely values free will. Individualism. Freedom. Liberty. The Right To Be Your Own Person.

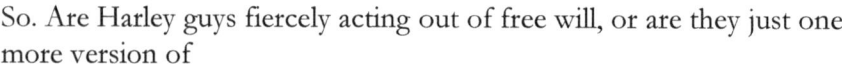

As much or much more maybe than any other subculture in the world.

But that's gotta be balanced with the even more fierce desire to belong, to fit in, to look right, act right, dress right.

So. Are Harley guys fiercely acting out of free will, or are they just one more version of

Pavlov's Dogs?

I'm gonna say this cuz I think it's pretty much true: the only people who are truly different from everyone else

tend to be certifiably nuts.

Because if *normal* people show up to their group of buds and they don't look like the rest of the buds, then the social pressure (as in, making serious fun of the idiotic way they look) is *huge* and they tend to show up next time fitting into the norms.

Even if they are all Harley guys.

Because. We all want to belong. To fit in. To be like the people we want to be like.

There are (I propose) two things in life that motivate us.

One. To belong. To be loved. To be a part of something that is greater than ourselves.

Two. To matter. To make a difference. To leave a mark. To leave the world a different (presumably better) place than it was. We want our lives to have a meaning and purpose.

And we will largely do whatever it takes to get those two things done.

Pavlov. Those are our stimuli. To be loved. To matter. And we will slobber with all the dogs to be loved and to matter.

There may be no free will about it. OK, maybe we choose which groups we want to be a part of, but I'm thinking that we are (mostly?) (completely?) influenced in that decision by the experiences we have.

For example. I have no desire to be a Harley guy or for my wife to be a Harley babe.

But I love to ski. I'm really good. So I want to look like a really good skier. I want to dress like one. I want to have skiis and boots and even poles that really good skiers might have. I want to have a fine-looking ski jacket, and fine- looking goggles, and a cool helmet, and cool gloves, and even (this is true) cool skier long-underwear.

Not cool. Totally not cool.

Now. If I have all of this cool gear and then I ski like a dork, well, that's not cool. Then I'm pretending to belong when I don't.

But I don't ski like a dork. So I get to belong. And my wife is a fine skier. So I have a beautiful ski wife. Awesome. It's the whole package. If I call her a beautiful ski babe or ski chick, I'm going to find one of my ski poles lodged in an uncomfortable place. So I'm not going to do that.

So here's the question. If life pulled a Tom Hanks "Castaway" trick on me and I ended up alone with a volleyball in a deserted ski resort, would I still ski? Am I making a free will decision to ski, or am I doing it because I want to belong to cool skiers?

Aye. That's a good question.

Because the reasons that I am a skier are 1) my parents took me skiing once in Colorado, and then 2) we moved to Switzerland when I was in

Curling groupies. Clearly nobody would look like this by making a free will decision to do so.

high school, and 3) skiing was The Thing To Do in Winter in Switzerland, and we had a ski team and I was on it (briefly) and so 4) skiing is totally cool. To me, anyway.

Curling to me is not cool. It's ridiculous. But to curlers, curling is awesome. There might even be curling groupies. Of course there are curling groupies. Clearly not a free will decision to be a curling groupie. Not a free will decision to write me the hostile curling groupie letters I'll be getting.

I'm fully convinced that curling was invented by drunk "people" (meaning drunk "guys") in the Far North who said to themselves one night, what can we do with ice, a big rock, and some brooms? And can we wear ridiculous pants, too? Awesome!

Anyway. So the question is, are ALL of our decisions caused by influences outside of ourselves, or just MOST of them? And if it's MOST, then where does free will actually start?

And the answer is...

Nobody really knows.

So everybody just makes stuff up.

Chapter 41	Free Will. Get over it. Or not.

So. Do we have free will or not? We (really really)(are totally compelled by the lack of free will to really really) want to know.

We should maybe read all of the very latest highly intelligent totally informed scientifically science stuff about it.

Here you go, then. From the Atlantic Monthly, June 2016:

"There's No Such Thing as Free Will

"The sciences have grown steadily bolder in their claim that all human behavior can be explained through the clockwork laws of cause and effect. This shift in perception is the continuation of an intellectual revolution that began about 150 years ago, when Charles Darwin first published On the Origin of Species.

"Shortly after Darwin put forth his theory of evolution, his cousin Sir Francis Galton began to draw out the implications: If we have evolved, then mental faculties like intelligence must be hereditary. But we use those faculties—which some people have to a greater degree than others—to make decisions.

"So our ability to choose our fate is not free, but depends on our biological inheritance.

"Galton launched a debate that raged throughout the 20th century over nature versus nurture. Are our actions the unfolding effect of our genetics? Or the outcome of what has been imprinted on us by the environment? Impressive evidence accumulated for the importance of each factor. Whether scientists supported one, the other, or a mix of both, they increasingly assumed that our deeds must be determined by something.

"In recent decades, research on the inner workings of the brain has helped to resolve the nature-nurture debate—and has dealt a further blow to the idea of free will. Brain scanners have enabled us to peer inside a living person's skull, revealing intricate networks of neurons and allowing scientists to reach broad agreement that these networks are shaped by both genes and environment.

"But there is also agreement in the scientific community that the firing of neurons determines not just some or most but all of our thoughts, hopes, memories, and dreams.

"We know that changes to brain chemistry can alter behavior—otherwise neither alcohol nor antipsychotics would have their desired effects. The same holds true for brain structure: Cases of ordinary adults becoming murderers or pedophiles after developing a brain tumor demonstrate how dependent we are on the physical properties of our gray stuff.

"Many scientists say that the American physiologist Benjamin Libet demonstrated in the 1980s that we have no free will. It was already known that electrical activity builds up in a person's brain before she, for example, moves her hand; Libet showed that this buildup occurs before the person consciously makes a decision to move. The conscious experience of deciding to act, which we usually associate with free will, appears to be an add-on, a post hoc reconstruction of events that occurs after the brain has already set the act in motion.

"The 20th-century nature-nurture debate prepared us to think of ourselves as shaped by influences beyond our control. But it left some room, at least in the popular imagination, for the possibility that we could overcome our circumstances or our genes to become the author of our own destiny. The challenge posed by neuroscience is more radical: It describes the brain as a physical system like any other, and suggests that we no more will it to operate in a particular way than we will our heart to beat. The contemporary scientific image of human behavior is one of neurons firing, causing other neurons to fire, causing our thoughts and deeds, in an unbroken chain that stretches back to our birth and beyond.

"In principle, we are therefore completely predictable.

"If we could understand any individual's brain architecture and chemistry well enough, we could, in theory, predict that individual's response to any given stimulus with 100 percent accuracy."

So there you go. No free will.

Except. But. There's always a but. From NewScientist, 6 August 2012:

"Brain might not stand in the way of free will

"Advocates of free will can rest easy, for now. A 30-year-old classic experiment that is often used to argue against free will might have been misinterpreted ... So what does this say about free will?

"If we are correct, then the Libet experiment does not count as evidence against the possibility of conscious will..."

Now, you might say, looking carefully at the dates, that one is 4 years before the other one. So doesn't that mess it all up?

No, it means that the Atlantic writer wasn't paying attention in 2012.

Science is so complicated.

And lo and behold. Check this out from ScienceDaily.com in July 2016:

"Johns Hopkins University researchers are the first to glimpse the human brain making a purely voluntary decision to act.

"Unlike most brain studies where scientists watch as people respond to cues or commands, Johns Hopkins researchers found a way to observe people's brain activity as they made choices entirely on their own.

"For the first time, researchers were able to see both what happens in a human brain the moment a free choice is made, and what happens during the lead-up to that decision -- how the brain behaves during the deliberation over whether to act.

"The actual switching of attention from one side to the other was closely linked to activity in the parietal lobe, near the back of the brain. The activity leading up to the choice -- that is, the period of deliberation -- occurred in the frontal cortex, in areas involved in reasoning and movement, and in the basal ganglia, regions deep within the brain that are responsible for a variety of motor control functions including the ability to start an action. The frontal-lobe activity began earlier than it would have if participants had been told to shift attention, clearly demonstrating that the brain was preparing a purely voluntary action rather than merely following an order.

"Together, the two brain regions make up the core components underlying the will to act, the authors concluded.

"'What's truly remarkable about this project,' said Leon Gmeindl, a research scientist at Johns Hopkins and lead author of the study, 'is that by devising a way to detect brain events that are otherwise invisible -- that is, a kind of high-tech 'mind reading' -- we uncovered important information about what may be the neural underpinnings of volition, or free will.'"

So once again, free will is back. Last month. Next month? Who can tell?

And finally, from the smartest man alive, physicist Ed Witten:

"I think consciousness will remain a mystery. Yes, that's what I tend to believe.

"I tend to think that the workings of the conscious brain will be elucidated to a large extent.

"Biologists and perhaps physicists will understand much better how the brain works.

"But why something that we call consciousness goes with those workings, I think that will remain mysterious.

"I have a much easier time imagining how we understand the Big Bang than I have imagining how we can understand consciousness...

"The brain, apparently the seat of free will, may forever remain a mystery. Or at least until everyone's forgotten about my having said so."

Chapter 42   Now It Gets Confusing

So. Here's what we are saying.

In a universe with a starting point where things are not always predetermined (because of quantum uncertainty) or predictable (because of Chaos Theory and the Butterfly Effect), then it is possible that both God and Free Will exist.

That's because thinking types back in the 18th Century said that if the universe is predetermined by the laws of physics and totally predictable and infinitely old, then

1) you don't need God to do anything and

2) you can't have free will.

The laws of physics make everything happen, they've always been there, nobody made them, nobody made anything, and they make you do everything you do.

But since the universe is not infinitely old and the laws have not always been there because they came into existence with Big Bang, and quantum uncertainty is the Way Things Work, then

1) you might need God to make everything happen and

2) you might have free will.

Nice. God and Free Will got thrown out together, so they get to come back together. Sweet.

However. There are some skeptical types who say things like this:

If God already knows everything that's going to happen, then you can't have free will, since he already knows what you are going to do.

That is...

if God exists, then free will can't.

Or, if God exists, then HE predetermines everything, not the

laws of physics.

Hm. That *is* a problem.

Because now we're in a place where free will only exists if God exists, but if God exists, then free will can't.

So what you gotta do is look at assumptions and conclusions and see if they are good physics. Since if God exists, he made physics that way that it is.

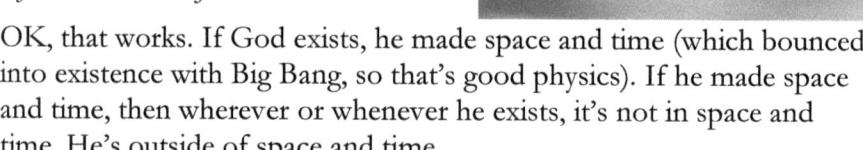

Here we go.

"If God knows the future"

OK, that works. If God exists, he made space and time (which bounced into existence with Big Bang, so that's good physics). If he made space and time, then wherever or whenever he exists, it's not in space and time. He's outside of space and time.

(Make sure you get this part. God doesn't exist in a where or a when. He does not occupy a point in space and/or time.

(That's the way we [who exist in points in space and time until we don't exist anymore] understand existence.

(So if someone asks the question, does God exist?, the answer is, what do you mean by existence?

(If he exists, he does not exist in the same way that we do. He created the way that we exist.)

Anyway.

Since he's God (if he exists), then, well, he's God, and he can see all of space and time together. Now since to a photon traveling at the speed of light, all of space and time are one point, then clearly God is at least as cool as a photon, so since he can see all of space and time, then he knows the past, the present and the future all at once. He sees it all. But not like a movie that he has already watched. Everything is a point to God, so he sees it all at once.

That's just physics.

"The future must already be determined, and if the future is already determined, we have no control over our future actions."

You wanna take a shot at this one? Go ahead. I'll wait.

(clock ticking. background humming. toe tapping.)

Time's up. Let's see how you did.

Problem the first: our cartoonist, and all the thinkers and philosophers who derived their thinking and philosophy from this cartoon (or maybe vice versa), is/are fooled by time. S/he isn't thinking about God outside of time, but God inside of time, as though the place God hangs out is just like the time and place we hang out, and, well,

it isn't.

We don't know what it is, but it's not time and space the way that we experience it.

So trying to catch God in a space-time trap isn't gonna work.

Problem the second: our cartoonist *et al* is blissfully unaware that this universe, the universe that if God exists and created it, is quantum. And it's chaotic.

And thus, nothing is predetermined. None of our lives, our decisions, our actions, anything we do is predetermined by anything except maybe behavioral conditioning, and we don't know where that starts and stops.

And that's the way God made it, because 1) if he exists 2) he made it and 3) intended for us to find it 4) the way that it is.

Now you might say, as my atheist friend Trevor once said, well, just because *we* can't know all of the forces that cause things to happen chaotically, doesn't mean that they can't be known by *someone* (in a God-like sort of way). And thus, things *are* predictable.

To which I responded, those forces might be capable of being known, but they get smaller and smaller until they become quantum, and then

they cannot be known anymore, because at the quantum level, nothing is predetermined. Ultimately, everything is quantum, and nothing is predetermined.

So here's the answer. God knows what you are going to do. No, that's not right. God knows what you did, what all of us did, because he sees all of time at once. But what you are going to do is not predetermined and is up to you.

He sees the beginning and the middle and end game all at once. But you get to ride the ride and make the decisions yourself.

If you like, it's a quantum thing. He both knows and does not know, depending upon how he is looking. It's all in the observation. And you, as the observer yourself, have a choice to do, or not to do.

Time is not seen by God as a movie. It is a photo. Even though you, and I, and everybody else lives it as a movie.

At least, that's what the physics of the universe tells us.

Problem the third: S/he (the cartoonist) has been fooled by an unfortunate misunderstanding of God's interaction with humans, created by well-meaning but sadly confused religious folk.

We'll talk about that next.

Chapter 43   God Does Not Have a Wonderful Plan for Your Life. Sorry.

A lot of well-meaning but sadly confused religious folk have been walking around thinking this:

God has a wonderful plan for your life.

And so. Our lives are predetermined, so it's not our fault if we do crappy things to each other, and so hell isn't really fair.

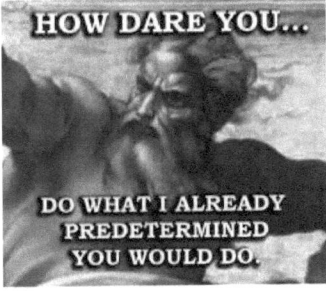

That is, God has this wonderful plan, but we, um, aren't interested since sex, drugs and rock&roll are so much more fun (or partying or getting drunk or being a hedgefund manager or a dictator or a serial killer or a rapist or a drug dealer or just your garden variety very nice kind to others giving and generous atheist), so we miss the plan and go to hell. Dang. Doesn't seem right. The last one, I mean.

But our religious folk are confused.

God does not have a wonderful plan for your life.

Your life is not predetermined by God.

Nobody's is.

You get to make decisions. And the decisions add up to being the life you create.

You can make God-ish decisions. Feed the hungry. Care for the homeless. Fight injustice and oppression. Don't be a jerk.

Or you can make not-God-ish decisions. Pretty much the opposite of all of those. Basically, be a jerk.

Religious folks somehow got the idea somewhere along the way

199

that God had everyone's life entirely planned out, start to finish, and you just had to figure out what his plan for you was, and then all would be peachy.

Only he doesn't tell you what the plan is.

Which is kinda nasty.

So you have to guess a lot and hope for the best. And pray. There's lots of praying.

So, the assumption goes, every minute is planned out. And you have to guess right. And he's not telling.

That's not the way it works. Nothing is predetermined. Nothing is planned out. Things might go well. They might go really really badly.

But. Regardless of what choices you make in your life, you're supposed to do the same thing all along the way.

Don't be a jerk. Love God. Love your neighbor. Be nice.

Now. Since God is God (if he exists) and can do anything he wants, he surely might have a plan for someone every now and then. There are examples.

But for most of us, and when I say most, I mean statistically pretty close to all of us, there's one plan.

Live your life. Make good choices. Do the right thing. Don't be a jerk.

That's free will. Take whatever job you want. Marry whomever you want. Live wherever you want. Do whatever you want to do. Live your life. Make good choices. Do the right thing. Don't be a jerk.

Here's what I'm telling you. God doesn't have a job, a spouse, a house, a louse or anything else planned out for you. I'm not saying he doesn't care, but it's not planned and waiting for you.

You get to choose. It's a free will thing.

> What is the secret of success? Right Decisions. How do you make right decisions? Experience. How do you get experience? Wrong decisions.
> relationshipquote.tumblr.com

If you assume that God has planned it all out, and all you have to do is figure it out via prayer and fasting, then there would be, it seems to me, only two types of decisions.

Perfect ones.

Crappy ones.

That's it.

But life doesn't work that way.

Because there are no perfect choices.

Jobs - every job is going to be crappy at some point or another.

Spouses - every spouse is imperfect, and that includes both sides of the spouse coin.

> EVERYTHING HAPPENS FOR A REASON. BUT SOMETIMES THE REASON IS THAT YOU'RE STUPID AND YOU MAKE BAD DECISIONS.

Living here, living there - there are no perfect places to live, and I'm a good source, because I've lived in Zurich, Geneva, Pebble Beach, Monterey, and Colorado, which are pretty danged spectacular places to live.

But not perfect. Spectacularly not perfect, in fact.

Maybe you think God wants you to go to Africa or the inner city or China.

It would surely be fine for you to go there and do whatever it is you think that you are supposed to do.

But there stands an excellent chance that it will surely not be perfect when you get there.

You'll be lonely. Bored. Lusty. Frustrated. Scared. You'll hate the locals and the local culture at some point, maybe at all points.

And if you think that you might not have done better to have stayed where you were, well, you easily might have.

Going, staying. Either is fine. Either is a good decision. Both come with good, bad and ugly parts.

It may be that the only real free will decision you get to make is kind of Hamlet-y. To be a jerk or not to be a jerk. Wherever you are and whatever you do.

That's worth thinking about.

BTW, I think it's useful to make the more challenging choice sometimes. Pain and suffering are good for you. Up to a point.

Chapter 44   Why bad things happen to good people,
               or Chaos Theory Explained

Here are some things folks worry about.

Like, why bad things happen to good people.

And, why good things happen to bad people.

Also, does everything really happen for a reason?

Plus, if a door closes, does a window open?

And most importantly, why do bad things happen to ME? Or in your case, YOU?

Now the reasons that we wonder about these things are curious.

It's because there's a certain ... tension ... in our thinking.

Science tells us that there's no reason or purpose for anything.

Not for the universe. Not for the earth. Not for humans.

Not for you. Not for me.

Everything is meaningless. That's what they tell us.

What? You don't believe me?

Well. Since science is all about evidence, here's some. Evidence, I mean.

Richard Dawkins, Oxford zoologist

...the universe "has precisely the properties we should expect if there is, at bottom, no design, no purpose, no evil and no good, nothing but pointless indifference"...human beings are "machines for propagating DNA".

Jacques Monod, Nobel Prize winner, physiology/medicine

"The ancient covenant is in pieces. Man knows at last that he is alone in the universe's unfeeling immensity, out of which he emerged only by chance."

Carl Sagan – Pale Blue Dot

"Look again at that dot. That's here. That's home. That's us. On it everyone you love, everyone you know, everyone you ever

heard of, every human being who ever was, lived out their lives. Our posturings, our imagined self-importance, the delusion that we have some privileged position in the Universe, are challenged by this point of pale light. Our planet is a lonely speck in the great enveloping cosmic dark. In our obscurity, in all this vastness, there is no hint that help will come from elsewhere to save us from ourselves."

Steven Weinberg, Nobel Prize winner, physics

"Though aware that there is nothing in the universe that suggests any purpose for humanity, one way that we can find a purpose is to study the universe by the methods of science, without consoling ourselves with fairy tales about its future, or about our own."

Bertrand Russell

"That man is the product of causes …;

that his origin, his growth, his hopes and fears, his loves and his beliefs, are but the outcome of accidental (collections) of atoms;

that no fire, no heroism, no intensity of thought and feeling, can preserve an individual life beyond the grave;

that all the labors of the ages, all the devotion, all the inspiration, all the noonday brightness of human genius, are destined to extinction in the vast death of the solar system,

and that the whole temple of man's achievement must inevitably be buried beneath the debris of a universe in ruins;

all these things, if not quite beyond dispute, are yet so nearly certain, that no philosophy which rejects them can hope to stand.

Only within the scaffolding of these truths, only on the firm foundation of unyielding despair, can the soul's habitation henceforth be safely built."

Stanley Kubrick: "The very meaninglessness of life forces man to create his own meaning."

Joseph Conrad: Life is "that mysterious arrangement of merciless logic for a futile purpose."

Jean Paul Sartre: "It is meaningless that we are born. It is meaningless that we die."

Sartre: "Every existing thing is born without reason, prolongs itself out of weakness, and dies by chance."

Sartre: "Everything has been figured out, except how to live."

Clarence Darrow: *Life is a ship that is "tossed by every wave and by every wind; a ship heading to no port and no harbor with no rudder, no compass, no pilot, simply floating for a time, then lost in the waves."*

The catch, of course, is that nobody lives like that.

We, and by "we" I mean "everybody", live like there's a reason for us to be here.

And thus, there's a reason why things happen. Bad things. Good things.

And there's supposed to be some sort of ... balance to the universe.

Justice.

Like, if you are bad, then bad things are what you get.

And if you are good, then you get good things.

The universe is supposed to be sort of like Santa Claus. There's a list. All of our names are on it. And good and bad things get divided up perfectly and justly and fairly between good and bad people.

Now, if there's no meaning to the universe, all of that is just ridiculous. Things just happen randomly without meaning or purpose, so when a bad or a good thing happens to you or me, well, tough luck or wasn't that nice?

So bad things will happen to good people and good things will happen to bad people randomly and without rhyme or reason, and good things will happen to good people and bad things will happen to bad people randomly and without rhyme or reason.

And sometimes more good things will happen to bad people and sometimes more bad things will happen to good people, because it's all random.

WHEN LIFE CLOSES A DOOR IT OPENS A WINDOW

...then zombies come in and eat you

functionally-insane.com

Nothing happens for a reason. Things just happen to people because things just happen and people are always getting in the way.

When a door closes, a door closes. Windows opening are entirely unrelated and random. Sometimes windows don't open when doors close.

In fact, if we really paid attention to Richard Dawkins up there, there are no good things or bad things, because there is no good or bad. There are just things.

"Good" and "bad" are just levels of comfort and inconvenience that we make up out of nothing. Because we persist in thinking that we matter. That we are here for a reason. That there is meaning and purpose to our lives.

Silly us.

Perhaps, though, we might be wrong. In part, at least.

That would be ... intriguing.

Chapter 45   Life is Chaotic. Then you die.

And when we say Chaotic, we don't mean messy (though life is certainly sometimes messy) but unpredictable. That's what Chaos Theory is all about. Unpredictability.

Life is unpredictable. And then, yes, well, it just so happens that at the end, dying ceases to be an option and becomes an imperative.

That might be the only predictable thing about life, in fact. Taxes too, I suppose. Unless you're able to figure out how to not pay them. As some have been known to do.

But just because life's unpredictable doesn't mean it's random.

Nor does it mean that there is no meaning or purpose to existence.

Or that there is no good or evil.

Richard Dawkins and his buds have mistaken unpredictability for randomness.

So the way that life works lies somewhere between randomness and predictability.

So when we say things like God has a wonderful plan for your life or when a door closes a window opens or everything happens for a reason or ask questions like why do bad things happen to good people or most importantly why do bad things happen to ME?,

we are acting like life is predictable. That there's a reason for everything.

That somebody is in charge and making all the decisions and pulling all the strings and making everything happen according to some grand plan.

God. Or if you don't like God, the Flying Spaghetti Monster or Pay No Attention to that Man Behind the Curtain or Fate or Destiny or Karma or Aliens or the Buddha or the Military-

Industrial Complex or the Illuminati.

We're gonna go with God. Short, sweet, simple.

Now the Richard Dawkinses of the world don't wanna go with God or anything resembling God, so they swing all the way to the other end of the bell curve to randomness.

Frankly, that just shows a lack of imagination.

Let's just use some. Imagination, I mean.

So. Let's just say that life will have good things and bad things in it. Also lots of things that are neither good nor bad but are just ... things.

Now. Everybody is going to get some good things and some bad things.

That seems fair.

So when something bad happens to us, here's what we say - why me?

Yeah, we never say that when good things happen.

We kinda assume that good things are normal and bad things are not normal.

For those who believe in God, sometimes we think that God's main job is to protect us from bad things and make sure we only get good things.

That's just silly.

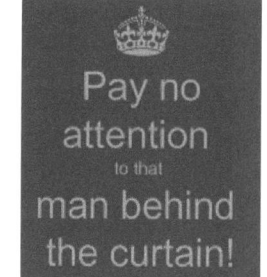

For those who don't believe in God or the Flying Spaghetti Monster or Pay No Attention to that Man Behind the Curtain or Fate or Destiny or Karma or Aliens or the Buddha or the Military-Industrial Complex or the Illuminati, bad things are just baaaad things dangit, and good things are great, unless you are Richard Dawkins pretending for a brief moment that there are no bad or good things, there are just things, in which case, a bad thing is just a thing.

Yeah, he never does that. If he ever did, he'd stop saying that religion was bad. For example. He's a total hypocrite. Sorry, Dicky. And he never ever ever ever says that a bad thing that happens to him is not a bad thing it's just a thing. Never ever ever. Nobody does, not even Dr. Dicky Dawkins.

So really, life is going to have good things and bad things in it.

Now. Here's a math question for you. Or two. Several.

Are the good and bad things going to be equally divided in your life? Just exactly the same number of good things as bad things?

And are they going to be equally divided between humans? That is, will every human get exactly the same number of good and bad things as every other human?

And will the good things and bad things happen at regular intervals so that we will know they are coming and 1) bake a cake (in case it's a good thing or 2) run and hide (in case, well, you know)?

And since Death is like the ultimate Bad Thing, is everybody going to die at the same age?

Whether you believe in God or not, the answer to all those questions is the same:

No. That's ridiculous.

So here's what that means. It means that you will have an unequal number of good and bad things happen to you in your life.

Some of you will get way more good things, and others will get way more bad things. Some won't suffer much. Others will suffer a crapload.

Sometimes in life you'll get a crapload of bad things all at once, and then there might be a looooong pause where nothing bad really happens. Sometimes a ton of good stuff will happen all at once.

And since we're all going to die at different times in different ways, some of us will die before we are born, some when we are born, some when we are babies or toddlers or pre-schoolers or schoolers or teenagers or collegians or young adults or middle adults or old adults, and some will die quickly and painlessly and others will not be that lucky.

And some things will happen for a reason, and others will not.

And sometimes doors just close and windows don't open. Sometimes they'll both open.

It's not random. It's unpredictable.

Your life is not paint-by-numbers. It's not fill-in-the-blank. It's not a crossword puzzle.

It's a blank canvas with dangers and joys lurking in the paint, the brush, in your fingers and brain, in the fabric of time and space and in the interactions that are your emerging life.

And every interaction is the artist of your life.

So here's what Chaos tells us - life is clumpy. Life comes in clumps. Clumps of good things. Clumps of bad things.

And when you get a clump of bad things that happen to you, you wonder, why do bad things happen to good people, but of course, what you really mean is, why do bad things happen to ME?! ALL THE DANG TIME!?!?

And the answer is (envelope, please), why not?

Bad things happen to everyone eventually. Sometimes terrible things.

But not equally. Some people have lots of terrible things happen to them.

And others ... don't.

Because the universe is clumpy.

In scientific terms, the universe started out not clumpy at all, and then got clumpier as time went on. So after 13.8 billion years, there's a crapload of clumpiness out there.

So just about the only other predictable thing about the universe is that it will be unpredictably clumpy.

Live accordingly. OK, that is not helpful advice.

Chapter 46   Living within the Clumps

Life is clumpy.

Get over it.

Not helpful.

Really. I mean, how are we supposed to live with unpredictability?

Well. Pretty much the way you've been living so far.

This shouldn't be a big surprise to you. That life is unpredictable, I mean.

Because, you know. Look around. See?

We really really want life to be predictable, so we say things like when a door closes a window opens and there's a reason for everything, but in fact,

that's just not true.

Here's what is true.

First. When a door closes, you can't go that way anymore.

So you've got to find another way.

Chances are good you'll find one. There are lots of options. Sometimes you need to be creative and work hard and be patient. It would have been good for you to get ready for a clump of bad things to come along, so, you know, having some extra money in the bank is a fine idea. For example.

If you know that bad clumps are always out there lurking, then heck. Get ready for them.

And everything doesn't happen for a reason, but lots and lots of things that happen come with a life lesson wrapped up inside, something to learn, something to remember for next time.

Like, wear a seatbelt. Flush twice. All smells are particulate. OK, that last one is not useful to remember at all. It's just disgusting.

However. There are some things that happen that are so terrible, horrific, and monstrous, that not only did they not happen for a reason, but they didn't happen so as to provide you with a neat little lesson to learn from.

They happened because the quality of bad things is clumpy, too. Many bad things are just kinda bad, maybe most bad things. But some are much, much worse. Some are beyond imagining. Some are the stuff of nightmares that are not supposed to transition into daytime and reality. But they do.

Some you just have to survive as best you can. And maybe find people who have already survived as best they could. And then maybe become one of those people, so that you can be one of them when it happens to someone else.

Could be that there is a reason, after all.

We are supposed to help each other. And sometimes it helps to have been there for people who are now there.

Community is the thing. Coming together. Sharing lives and pains and joys and struggles and clumps.

Oh. And sometimes bad things happen because you did something stupid.

Let's not rule that out.

Because that's when everything does happen for a reason, and the reason is, you were an idiot.

Now. If God exists (always an option), then how does he fit into all of this?

Well. His main job is not to keep you safe from bad things. If it is, then he's doing a really crappy job of it.

And we're going to assume that God never does a crappy job on anything.

His main job is not to provide you with object lessons to learn from. With an exception for the stupid things you do. Those, you're supposed to learn from, whether God exists or not. Here's the lesson: That thing you just did? That's what idiots do. Don't be an idiot.

See how easy it is?

Now, as far as all the other bad things, especially the really bad things that happen, that's an interesting question.

Clearly if God exists, he could stop bad things from happening.

And clearly, he doesn't always do that. He might do it sometimes, but, well, you never really know. 'Cause they didn't happen.

So why doesn't he just stop all the bad things and keep only the good things?

Huh. Good question.

Let's give it a shot.

First. If there were no bad things, then we would never have any reason to look to God for help. Or to each other, for that matter.

We might even just start to think about only ourselves all the time, and never about anybody else.

Eventually, that would cause bad things. Maybe it does already. I'm gonna bet on that one.

Second. if there were no bad things, then there would be different levels of good things. And we'd start to rank them so that some of them would start to seem bad to us, and then we'd complain about it.

So even if there were no bad things, we'd come up with some. Maybe we already do that. I'm gonna bet on that one, too.

Third. There are only good and bad things if there is some sort of God around to call them good and bad. Even Richard Dawkins says that. "No design, no purpose, no good and no evil, nothing but pointless indifference."

That's a universe without God.

So even if we're not totally crazy about all the bad things,

they are only bad if God exists.

So you just have to live with that. And you kinda have to live with whatever he calls good and/or bad. You don't get to get to define good and bad for yourself.

213

Because often something that is good for you, is bad for someone else. One of you is being a jerk.

And we can't really let nations or cultures define what is good and bad, because they are full of people, and people are often not the best judges of that.

Because, well, the Holocaust. For example. There might be others. Ha. Not funny.

Fourth. If there were no bad things, then we'd never do anything at all.

Because some of the bad things are like, um, being hungry. Thirsty. Tired. Dirty. Naked. Walking instead of driving. Or floating. Or flying. Being sick. Or lonely. Or bored.

All of those needs give us something to do. For ourselves. For each other. If we had no needs, we wouldn't do anything. We wouldn't need to. And we wouldn't need each other. Or God.

And we wouldn't have dark chocolate. Or ice cream. Cookies. Hamburgers. Pizza. Fondue. Ceviche. Curry. Sushi. Burritos.

Or cars or movies or jeans or books or bikes or skiis or games or TV or music or scuba diving or dancing or theater or pyramids or great walls or tall buildings or short buildings or buildings or

anything. At all. Like friends. Lovers. Children.

Bottom line. Life is clumpy. So we need each other. And we need God.

Get over it.

Chapter 47   Something Else Going On.

Having now pissed off all the religious people and all the Harley guys (and the latter is probably safer, since they'll just beat me to death with the Harley tool of choice, whereas the religious folk tend to have all sorts of interesting solutions to the problem of what they might define as heresy.) and probably the atheists and agnostics, too (do they have a plan for dealing with, um, what we might call anti-heresy? a-heresy? a-heretics?) ...

Anyway. Let's piss off the scientists.

Some of them, anyway. The old school traditionalists.

Here we go.

Several chapters ago (you should go look - it's the Free Will - Get Over It. Or not. chapter. Chapter 41), there were some articles.

The first one said that free will, just like everything else, has to have evolutionary origins and therefore is just a genetic byproduct, just the DNA that each one of us inherited from all of our ancestors. Everything biological is a product of evolution, and that's true for the brain, and that's where free will would be, and so all of our decisions are predetermined by our genetics, and so we don't have free will.

And there were all these experiments that seem to say that we don't have free will.

Until in the second article, there was an even better experiment using an MRI that actually observed free will apparently happening in the brain.

And then, of course, in the final article, there is the reality that we'll never really understand the brain, anyway. So we're never really going to know whether or not we have free will based on any kind of scientific evidence.

I read something else about that the other day - someone said that if the brain were simple enough for us to understand it, we'd be too simple to understand it.

Zing.

Let's assume for a minute that the absolutely positively latest and best experiment, the one that says we have free will based on the MRI evidence, let's assume that one is correct.

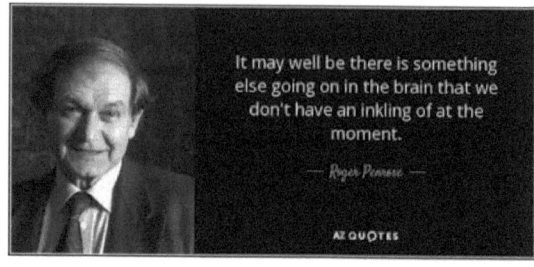

Let's assume that we have free will.

Standard Darwinian evolutionary theory says we can't have free will. Neo-Darwinian evolutionary theory says we can't have free will.

But, today at least, we do.

Hmmm. That is a conundrum.

So either evolutionary theory the way that we understand it is generally wrong, or specifically wrong.

Let's assume that it's generally right, but specifically wrong. Like, in this case.

So what does that mean?

Well. Regular old evolutionary theory says that random mutation provides a gene by complete accident that happens at a certain moment in time and space to provide for better survivability, and that gene is then selected for, and so it goes.

What if there's something else going on?

You should keep reading now even if you're upset.

Well. Here's part of an article from NewScientist for you to read and consider:

"In recent years, our understanding of biology has taken huge strides. Advances in genetics, epigenetics and developmental biology challenge us to think anew about the relationship between genes, organisms and the environment, with implications for the origins of diversity and the direction and speed of evolution.

"In particular, new findings undermine the idea, encapsulated by the 'selfish gene' metaphor, that genes are in the driving seat. Instead, they suggest that organisms play active, constructive roles in their own development and that of their descendants, so that they impose direction on evolution.

"Some biologists are trying to shoehorn the new knowledge into traditional evolutionary thinking. Others, myself included, believe a more radical approach may be required. We don't deny the roles of genetic inheritance and natural selection, but think we should look at evolution in a markedly different way.

"It is time for the theory of evolution to evolve.

"We now know that things other than genes are transmitted from parents to offspring...

"These and many other findings suggest that the current focus on genetic mutations only captures part of the story of adaptive evolution – the slowly changing part. The broader view shows there are other ways to generate heritable variety.

"And that's not all. We now also know that a given set of genes has the potential to produce a variety of phenotypes, depending on the environment in which the organism develops.

"This ability, called developmental plasticity, used to be dismissed as "noise" or mere "fine-tuning", but recent research suggests it may

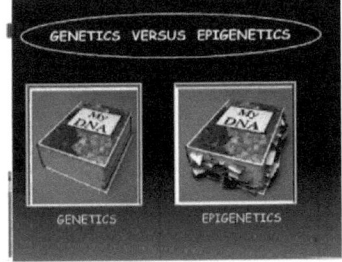

play a far more active role in the evolutionary process. As well as being able to respond in specific ways to particular conditions, organisms seem to have evolved the ability to respond flexibly to whatever conditions they experience...

"This allows systems such as the immune system, nervous system and behavioural systems (through learning) to adjust to meet whatever environment the individual faces.

"Perhaps, rather than merely setting limits on what forms are available for selection, developmental bias directs evolution by generating the tramlines along which the engine of selection can proceed."

The article goes on. You can read it here - www.facebook.com/www.lifeuniverseverything.org/posts/10154012596 767428 or if you have a subscription, at NewScientist.com.

So. Here's what we're suggesting.

Free will is not a product of random mutation and natural selection.

It's a product of directed evolution.

Now that is really gonna make absolutely everybody upset.

You need to remember, though, that if there is no free will, then you can't be upset at me because my DNA made me do it.

And if there is free will, then you can't be upset at me because, heck, there is free will.

Trapped like rats. Don't you hate that?

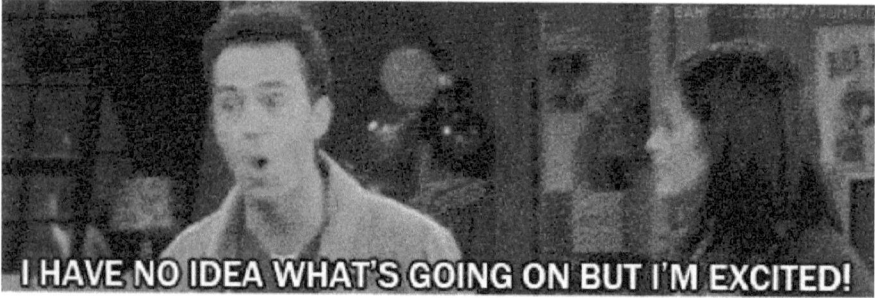

Chapter 48   Now we're talking about evolution.
              Not a great move, frankly, book-popularity-wise.

Yeah, writing about evolution in a blog is a great way to 1) attract trolls and 2) flamings and 3) fire- and brimstoning.

Because if you don't write exactly and precisely what everyone wants to read, then they'll come after you with frontal lobotomies and exorcisms and excommunications and really bad language. And they'll say you're stupid. All of them will say that.

Like, it's either TOTALLY RIGHT or it's TOTALLY WRONG.

There are no other options.

So just for fun, let's agree with the article we talked about last chapter from NewScientist magazine that evolution is in deep need of a fixin'. It has a problem. An issue. It needs to talk to someone about it.

The problem (spoiler alert!) is

Randomness.

Let's just also say that there's evidence for evolutionary change to be found.

Lots. Tons.

But let's also say that the assumption that random mutation is 1) the only thing making it happen and is 2) truly random, is 3) just an assumption that we've made up all the way along the way.

randomness

(noun)

The property of all possible outcomes being equally likely.

It was not Darwin's assumption. He came up with Natural Selection, but not Random Mutation. They didn't have genes then (everybody wore khakis)(sorry, bad joke). Genes were later. Mutation was later. Randomness was later.

Random mutation and natural selection are the way that neo-Darwinian evolution is thought to have always worked. Genetic mutation is random. Natural selection is not.

And OH. BTW. Remember how the thinkers got rid of God (and free will) by using an infinite universe? Which turned out not to be true?

Randomness in Evolution is how the thinkers get rid of God all over again.

That is, if everything is random and nothing happens on purpose and evolution is not heading anywhere ever, especially not towards humans, then

religion is just wrong to think that man is special. Because humankind, just like everything else, is just a big biological accident.

So. If Randomness is wrong (or mostly or partly or sometimes or frequently wrong), then humans may not be an accident and evolution was indeed heading somewhere and humans may be the where. Maybe. Along with free will. Maybe. And God might exist and might have structured the evolutionary imperatives as, well, imperatives.

You might go back and read the previous chapter again. So that you'll know where we're going.

Here's a hint: *We now know that things other than genes are transmitted from parents to offspring...*

That means that something other than random mutation and natural selection is going on.

That, frankly, is earth-shattering news.

Here's another bit: *As well as being able to respond in specific ways to particular conditions, organisms seem to have evolved the ability to respond flexibly to whatever conditions they experience...*

This allows systems such as the immune system, nervous system and behavioural systems (through learning) to adjust to meet whatever environment the individual faces.

And finally: *... developmental bias directs evolution ...*

So here's what it's saying.

It's saying that rather than, or maybe in addition to random mutation, organisms change deliberately and intentionally to things going on around them.

!!!!!!!

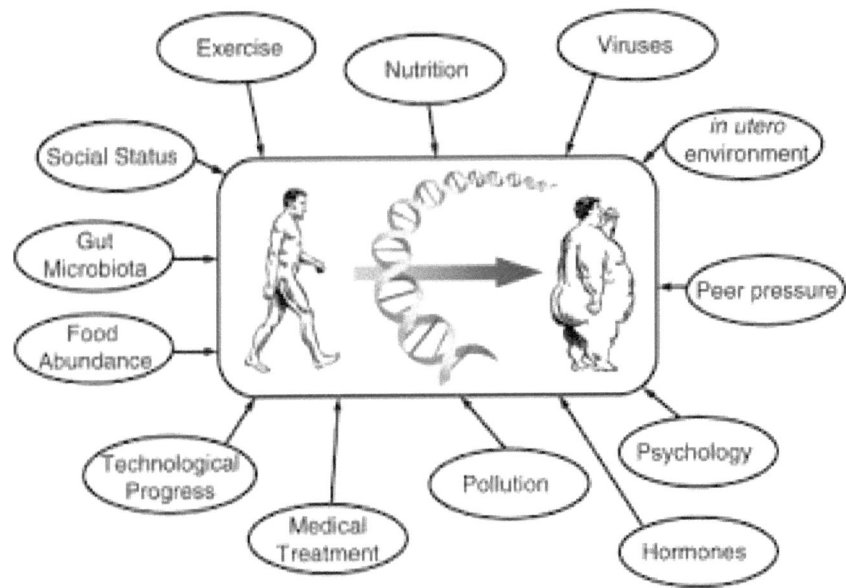

So evolutionary change is not always, and maybe never is or was random.

It may be sometimes, maybe most times, maybe all times intentional and organized. Specific. As though the organisms responded with deliberate intelligence to figure out what was going on and what they should do about it so that they don't die.

For the religious among the readers, let me tell what this is not. It is not intelligent design. It is not all-at-once miraculous creation. If there is anything intelligent design-y or miraculous-y about it, it is that organisms seem to have an innate, unexpected, surprising, and on the face of it miraculous ability to look around, see what's going on, and do something creative and amazing and generally entirely unexpected and unpredictable to fix it.

For the irreligious, areligious, atheist agnostic skeptical Newtonian readers out there, it is not random and it is not completely or maybe even mostly genetic. It is that via epigenetics, developmental bias, symbiotics, and/or spontaneous emergent self-organization, organisms seem to have an innate, unexpected, surprising, and on the face of it miraculous ability to look around, see what's going on, and do something creative and amazing and generally entirely unexpected and unpredictable to fix it.

It seems positively 1) neo-Larmarckian and 2) heretical.

Lamarck was the guy who suggested that organisms did this kind of thing.

All the trolls of his time made fun of him. Then he died. Not because of everybody making fun of him, but still, he's not around to get to say "I told you so." Bummer.

Because Lamarck is kinda back.

Evidence would be good.

Here's some, from www.evolutionnews.org :

"Jean-Baptiste Lamarck (1744-1829) was an early evolutionist who proposed that life forms could acquire information from their environment and pass it on in their genes. He was dismissed, when not ridiculed, by Darwinists for many decades (though not, as it happens, by Darwin). But the basic thrust of his idea has recently resurfaced in epigenetics.

"Epigenetics is the study of the systems and processes by which genes' expression can be altered, not randomly as in Darwinism, but by specific, predictable, repeatable, and researchable events -- and then inherited in the altered state."

Here's another: "Science Magazine called Michael Skinner 'the epigenetics heretic' for maintaining that chemicals can cause changes in gene expression in mice that persist across generations. Notice who ... had the biggest knee-jerk reaction of all:

"Michael Skinner is gleefully listing the disciplines that he's ruffled with his contention that, without altering the sequence of DNA, certain chemicals can cause harmful health effects that pass down generations. Toxicologists are so outraged that they have tried to block his funding, he says. Geneticists resist having their decades-old understanding of inheritance overturned. Then there are the evolutionary biologists, who have 'the biggest knee-jerk reaction of all.'

"Skepticism is to be expected, Skinner acknowledges: 'This is probably going to be the **biggest paradigm shift in science in recent history**,' he declares." (Emphasis added.)

And that's just epigenetics. Wait till we get to Complexity Theory.

And from MIT. MIT. The real MIT. The IT that's in M.

"The effects of an animal's environment during adolescence can be passed down to future offspring, according to two new studies. If applicable to humans, the research, done on rodents, suggests that the impact of both childhood education and early abuse could span generations. The findings provide support for a 200-year-old theory of evolution that has been largely dismissed: Lamarckian evolution, which states that acquired characteristics can be passed on to offspring."

Just to beat it to death, three more studies from 2017.

From ScienceDaily:

"Today, there is a new reason why an expectant mother should put down that glass of wine -- drinking alcohol during pregnancy will not only affect her unborn child, but may also impact brain development and lead to adverse outcomes in her future grand- and even great-grandchildren."

And if you thought you'd bail out from booze to tea while preggers, check this out from Science Bulletin:

"…researchers show that tea consumption in women leads to epigenetic changes in genes that are known to interact with cancer and estrogen metabolism. The results are published in the journal Human Molecular Genetics." Tea. Only in women. It's not known if this is a bad or a good thing, yet.

And again from ScienceDaily in 2017:

"It has long been thought that these epigenetic modifications never cross the border of generations. Scientists assumed that epigenetic memory accumulated throughout life is entirely cleared during the development of sperms and egg cells. Just recently a handful of studies stirred the scientific community by showing that epigenetic marks indeed can be transmitted over generations … 'Our study indicates that we inherit more than just genes from our parents. It seems to be that we also get a fine-tuned as well as important gene regulation machinery that can be influenced by our environment and individual lifestyle. These insights can provide new ground for the observation that at least in some cases acquired environmental adaptations can be passed over the germ line to our offspring',"

If you're religious, you need to keep reading. If you're not, you need to keep reading.

Chapter 49   Evolving Evolution

Now there's a boatload of negativity about all of this. That's because there's been so much hostility about evolution that some folks are afraid to question it at all, and other folks are afraid that it'll turn out to be true, only in a different way than everyone thinks, OR, and this is true on both sides of the argument,

they're just not paying attention.

Both sides are stuck in arguing about Old Evolution, and they're not even aware that New Evolution is not only out of diapers, it's about ready to go to high school.

Neither side is ready for the argument to change in a dramatic, paradigm-shattered sort of way.

Curious, that.

Anyway.

Remarkably, the place where Lamarck was wrong was in the timing.

He thought (like Darwin) that changes in response to environment would take place over many thousands or millions of years.

Turns out, those changes can happen in one generation.

One.

Mother to child. And hence to grandchildren. Whether or not it continues further remains to be seen. And since we don't really know what's going on yet, well, it remains to be seen.

The Old School Traditionalists are all about "Gradualism", which is that things evolve gradually over enooooormous amounts of time. Slooooowly. Tiny little random mutations that add up to big changes but it takes a looooong time.

So that's Darwin and Larmarck and Richard Dawkins (who is reputedly still alive) and all the high school and college and university textbooks. Mostly.

Then there's the slightly Newer School guys like Stephen Jay Gould (who is sadly no longer alive. I mean, sadly for his friends and family.

Hard to know how he feels about it.) who had enooooormous fights with Dawkins about Gradualism, because Gould Didn't Think So.

He said that things went along pretty much as they were for looooong periods of time, and then kinda all at once, evolution happened in a much much shorter amount of time. Fast.

He called it "punctuated equilibrium", which is a fancy scientific way of saying that nothing happened for a looooong time until it was punctuated by big changes in a much much shorter amount of time.

He and Dawkins had big fights. Then Gould died. Then everybody sort of decided that the fossil record was much more about punctuated equilibrium and Gould was right after all.

Dawkins moped about and complained, but he was wrong and Gould was right.

Now, nobody had much of an explanation as to how things which were supposed to happen sloooowly didn't. Happen sloooowly, that is.

But now we kinda do. To review:

"As well as being able to respond in specific ways to particular conditions, organisms seem to have evolved the ability to respond flexibly to whatever conditions they experience...

"This allows systems such as the immune system, nervous system and behavioural systems (through learning) to adjust to meet whatever environment the individual faces.

"And finally: ... developmental bias directs evolution ..."

So NOW there's developmental bias and epigenetics and symbiosis and Complexity Theory.

Complexity Theory probably says it best. It says that using self-organization and spontaneous emergence, organisms and systems of organisms solve problems of survival with intent and specificity.

Not randomly. Not accidentally. Not always even genetically. In fact, it could be even mostly not always genetically.

Deliberately and on-purpose.

The universe organizes itself spontaneously.

That is, emergent self-organization is at the root of existence.

What emerges is not only more than, but vastly different from just the parts.

Self-organization results in higher forms of order.

But the process is unpredictable – you don't know what you'll get until you get it.

And problems are solved with intent and specificity.

Here's an example. It's about locusts, from LiveScience.com in about 2006.

"Scientists have finally figured out the exact moment when a jumbled swarm of creatures becomes an organized, unified, and sometimes terrifying, mass.

"Examining a group of desert locusts, researchers found that at low densities, the insects were unorganized and went their separate ways. But when the group's density increased, the bugs fell into an orderly line and began to follow the same direction.

"When there were a few of them together, they did not coalesce. As the group grew to 10 to 25 members, the locusts got closer to each other, but still did not move in unison.

"It was only when the researchers placed about 30 locusts in the arena that the insects fell into a line and started moving in the same direction.

"The march of the locusts is a bit of a mystery since they have no leader and each one can only communicate with close neighbors."

When the locusts reached a certain density, a "tipping point", then the self-organization emerged from that density. Not before. Individual locusts don't do this. It's not a genetic thing.

And what we found was something much more than just the sum of the parts, something dramatically different from the parts, almost like a super-organism, one massive locust monster from the deep.

The last sentence in the article is key. It's *a bit of a mystery*.

We don't know yet how it all happens.

We should probably find some more examples. What a fine idea.

Chapter 50 Not Your Grandmother's Evolution

I changed my mind.

I just found another bit on evolution and epigenetics from Michael Skinner in *Aeon.com*, who's getting to be my fave. My BSF. That's Best Scientist Forever.

And here're the best bits:

"The unifying theme for much of modern biology is based on Charles Darwin's theory of evolution, the process of natural selection by which nature selects the fittest, best-adapted organisms to reproduce, multiply and survive.

"The process is also called adaptation, and traits most likely to help an individual survive are considered adaptive. As organisms change and new variants thrive, species emerge and evolve.

"But this explanation for evolution turns out to be incomplete, suggesting that other molecular mechanisms also play a role in how species evolve.

"Part of the explanation can be found in some concepts that Jean-Baptiste Lamarck proposed 50 years before Darwin published his work. Lamarck's theory, long relegated to the dustbin of science, held, among other things, 'that the environment can directly alter traits, which are then inherited by generations to come'.

"Lamarck, a professor of invertebrate zoology at the National Museum of Natural History in Paris, studied many organisms including insects and worms in the late 18th and early 19th centuries. He introduced the words 'biology' and 'invertebrate' into the scientific lexicon, and wrote books on biology, invertebrates and evolution. Despite this significant academic career, Lamarck antagonised many of his contemporaries and 200 years of scientists with his blasphemous evolutionary ideas.

"At the start, Lamarck might have been pilloried as a religious heretic, but in modern times, it is the orthodoxy of science – and especially

Darwin's untouchable theory of evolution – that has caused his name to be treated as a joke. Yet by the end of his career, Darwin himself had come around; even without the benefit of molecular biology, he could see that random changes were not fast enough to support his theory in full.

"(Here) is the precise definition of epigenetics: the molecular factors that regulate how DNA functions and what genes are turned on or off, independent of the DNA sequence itself.

"Epigenetics involves a number of molecular processes that can dramatically influence the activity of the genome without altering the sequence of DNA in the genes themselves.

"Environmentally induced epigenetic transgenerational inheritance has now been observed in plants, insects, fish, birds, rodents, pigs and humans. The epigenetic transgenerational inheritance of phenotypic trait variation and disease has been shown to occur across a span of at least 10 generations in most organisms, with the most extensive studies done in plants for hundreds of generations.

"One example in plants, a heat-induced flowering trait first observed by Carl Linnaeus in the 18th century, was later found to be due to a DNA methylation modification that occurred in the initial plant, and has been maintained for 100 generations.

"In worms, traits altered by changes in nutrition have been shown to propagate over 50 generations. In mammals with longer generation times, we have found toxicant-induced abnormal traits propagated for nearly 10 generations. In most of these studies, the transgenerational traits do not degenerate but continue.

"Much as Lamarck suggested, changes in the environment literally alter our biology. And even in the absence of continued exposure, the altered biology, expressed as traits or in the form of disease, is transmitted from one generation to the next."

Here's what evolutionary folks did. They used Randomness in evolution to get rid of the need for God.

Here's what religious folks did. They missed that entirely mostly and decided that the earth and the universe were only 6000 or so years old and everything happened instantaneously, more or less, spread out over six days.

Here's where we get to meet in the middle. Evolution is a thing, but it is not a random thing. Things self-organize to solve problems of survival with intent and specificity. It's an intelligent process of reacting to the environment quickly and deliberately, not over long periods of time gradually and randomly.

It happens in ONE generation. And can persist for A HUNDRED generations.

Not gradually over long periods of time. Not randomly in any sense.

So since Randomness has gone away,

God is back.

Now. You might be saying, what a minute! Just because Randomness is gone doesn't mean God is back! WTH!

Well. Yes, it does. If they used Randomness to get rid of the need for God, then logically if Randomness is no longer Randomness, then it's pretty reasonable to say that God is needed. That's just logic.

But. Maybe you need a bit more than that. OK.

Evolution as a theory doesn't get rid of God. Just like the good ol' laws of nature, if the universe has a starting point, then the question, where do the laws come from, has as a possible and reasonable answer, God made the laws.

It's not the only reasonable answer, but, well, there are no other answers to the question, so you've got that or nothing.

You could then say, well, that's just God-of-the-Gaps all over again. Don't know the answer, so just blame it on God.

The problem with that is that God-of-the-Gaps is an inside-the-universe thing. Not an outside-the-universe thing.

Everything inside the universe can be explained (eventually probably) by the laws of physics.

But the laws of physics only exist inside the universe. Including, as far as we know, evolution. Which is really a process more than a law.

The universe produced the laws of physics, and evolution is apparently derived from those, and all of that happened because of Big Bang creating the universe itself, and so we have no laws to cause Big Bang and we've no real idea at all why there are laws and particles and forces and all of that. Why there is something rather than nothing, that is.

The laws of physics are intelligent and ordered and extraordinarily calibrated to the finest of fine-tunings, and if there is only one universe (which is all we will ever ever ever have evidence for), then, well, somebody smart set it all up. Even if there are many universes, there still could be somebody smart out there in the Nothing setting it all up.

Somebody out there somewhere.

For lack of a better word, we'll call that God.

So here's what we're saying. If things happen in life, the universe and everything (Nice phrase, that. I should remember it.) with clear and unmistakable signs of an intelligence, one that empowered the universe with the laws of physics and with the amazing power of epigenetically driven evolution driving things so that intelligent observers would arrive in the universe and cause REALITY ITSELF to come into being, then, well,

that intelligence, that organizing observing intelligence outside of time and space that observed the universe into being and imbued it with organizing laws and rules so that

HOMERSAPIEN

first there was the language of the universe (mathematics) and second there was physics and third there was physical chemistry (inside of stars) and fourth there was then chemistry (the table of elements) and fifth there was biochemistry (life itself) and sixth there were eventually observers and seventh there was Reality, then

you don't have to call it the God of Christianity or Judaism or Islam or the gods of Hinduism or the pantheonic gods of Rome or Greece,

but whatever. It's a lot like whatever God would be if God were.

And what we've already seen in previous posts is that the universe is apparently defined by

Interactions. *Relata*. Relationships.

Which means that this God is likely to be an interactional, relational God, and thus it is entirely reasonable to suggest that this God is all about relationships.

Perhaps even and especially with us. Dang. That's some sweet logic.

Chapter 51 What's It All About?

We maybe should probably define what "relationship" means. Or give it a shot or something. It might matter.

We've already talked about it in terms of physics.

Gravity is an interaction between matter and space-time.

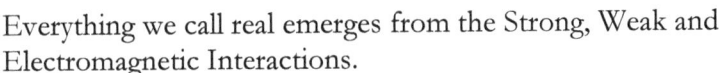

Space-time itself may be an interaction born of quantum entanglement.

Everything we call real emerges from the Strong, Weak and Electromagnetic Interactions.

There's the Higgs Interaction that keeps the universe from collapsing, provides for matter over anti-matter, and gives mass to critical particles.

Energy and matter are a quantum interaction via the Higgs. Big Bang itself arrives by virtue of a quantum interaction.

Dark Energy and Matter were only "discovered" because of their interaction gravitationally with space-time and galaxies, and those "discoveries" were only possible because we learned how to detect those interactions by interacting with energy waves. That's a fancy way of saying we figured out how to look at and measure them.

The chemistry of the universe arrives via the life and death of burning stars via the Weak Interaction, the stars themselves birthed and slaughtered by Gravity.

Chemistry itself is an atomic interaction, and biochemistry is an interaction from the very earliest, simplest living things all the way up through the tree of life (Yeah, "tree" doesn't work anymore. "Bush of life" is better) to the most complex of organisms, which somehow includes both politicians and reality show hosts. Not to mention reality show contestants.

Evolution is an interaction between organisms and the environment, between stasis and challenge, between genetics and epigenetics, between genes and the biochemical environment in which they find themselves.

And so on. Plus the things themselves don't exist except as part of an interaction.

Relationships may be all that actually do exist.

But there's a deeper level of interaction that we've also talked about, but need to talk about more. It's a Complexity Thing.

First, things only work because of a web of interactions.

Second, what emerges from the interactions is different than the individual players, the bits.

Here's a nice summary again from Chapter 49:

The universe organizes itself spontaneously.

That is, <u>emergent</u> <u>self-organization</u> is at the root of existence.

What emerges is not only more than, but vastly different from just the parts.

Self-organization results in higher forms of order.

But the process is unpredictable – you don't know what you'll get until you get it.

And organisms and systems of organisms evolve by solving problems with intent and specificity, not only and maybe not even primarily via random mutation and natural selection.

How, you might ask quizzically with bemused wonderment on your face, does this work when the universe is entropic, that is, it gets less complicated rather than more complicated?

A question that keeps you up at night, I'm sure. As it does all of us.

Well. "…entropy, a concept commonly conflated with disorder, can actually organize things.

"…simulations showed that entropy drives simple pyramidal shapes called tetrahedra to spontaneously assemble into a quasicrystal—a spatial pattern so complex that it never exactly repeats.

"The discovery was the first indication of the powerful, paradoxical role that entropy plays in the emergence of complexity and order."

I found that in Wired in 2017. It's danged good.

I found this on ScienceDaily in 2008:

"'Complex systems science is just the evolution of science.' Since the revolution that Newton and Descartes helped launch, the main thrust of so-called normal science has been to look for smaller pieces and more fundamental laws. Molecules yield atoms yield quarks.

"There are many problems that we figured out by breaking things into little pieces. Scientists figured out DNA with its double helix. And then they figured out the human genome by measuring like crazy. There was a sense conveyed that once we understood all the bits of the genome, we'd understand everything human, but that's totally insane."

"It's like saying once we understand atoms we understand matter. But we don't.

"Of course, many of the underlying ideas behind complex systems are far older than the name. It was Aristotle who stated that the 'whole is more than the sum of the parts.'

"But complex systems science takes this realization further. As physicist PW Anderson wrote in a 1972 paper in Science, in a complex system the whole becomes not only more than, but very different from the sum of its parts."

For example, take water. This is from LiveScience:

"Water makes up 70 percent of the Earth's surface and is the main component -- about 80 percent -- of all living things. But it is far from ordinary.

"We think we understand everything there is about a single water molecule ... What we don't understand so well is how they interact with each other."

And space-time and black holes. I don't know where I found this. Wait. Yes, I do. *Nature*, 2014:

"So putting it all together, it seems that entanglement is somehow related to space, and that computational complexity is somehow related to time.

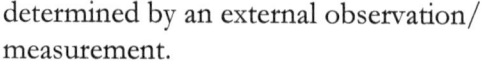

"Inside a black hole, 'Things fall because there is a tendency toward complexity ... the black hole's interior is protected by an armour of computational complexity'."

I don't really know what that means, but then, neither did whoever wrote it. The double use of the word "somehow" is a dead giveaway.

"Entanglement" is the ultimate quantum interaction – two particles are so much halves of the same quantum coin that their relationship happens across vast distances of time and space, and what happens is determined by an external observation/measurement.

An Interaction. Two interactions.

You look. They talk. Somehow.

Complexity tales are legion. Parasites control the brains and behavior of their hosts, even manipulating the way the bodies look in specific and intentional ways. The stories are headlined things like Zombie Ants and Disco Zombie Snails. Really. You should look those two up. They're a bit terrifying.

Maybe most terrifying is the narrative of the parasite called Toxoplasma Gondii.

T. Gondii lives, and can only live, in the intestine of cats, from which it gets pooped out. Since it needs to get back into the cat, it has a problem.

The first part of the problem is solved when a rat or a mouse eats the cat poop.

Can I just point out that rats and mice are not all that discriminating as gourmands?

So now the parasite is inside the mouse. How to get into the cat? How how how?

The next part of the problem is that lots of things like to eat mice. Dogs, foxes, wolves. Owls, eagles, hawks. Snakes. Drunk kids at frat parties.

But the parasite has to get into a cat.

So. It changes the mouse's brain with intent and specificity to do one thing.

It causes the rodent to lose its fear of cats.

Seriously.

Then it does another thing to the mouse. It causes it to be sexually attracted to 1) cats and 2) other mice of the opposite mouse gender that are infected with the parasite.

OK, that's two things.

So that the mouse is far more likely to end up in a cat than in anything else.

It changes the mouse's brain. Really.

So then they did a couple of studies.

Scientists did. Not the mice or the cats or the parasites. They don't do studies. Except for maybe the parasites, which seem to be running the show.

One of the studies was to find out when cats became popular in the modern era. The other study, separate and unrelated, was to find out when schizophrenia became common in the modern era.

The answer to both studies was the same. It was in Victorian England.

So now they (the scientists again) kinda think that maybe since 30% of the world's population seems to be infected with Toxoplasma Gondii, and since there's no cure, that things like schizophrenia, depression, self-harming, and even car accidents might have roots in the infection.

Here's what's going on at the very least – a parasite manipulates the brain of its host with intent and specificity to make sure that it ends up in a cat intestine. The different parasite that creates zombie ants and snails is working to get inside a bird by having the bird eat the ants and the eye stalks of the snails.

It's a parasite that is engaging in a complex interaction with its host to fool a cat or a bird into eating the host.

Chapter 52   The Strange Case of the Talking Bacteria

Bacteria do this, too. From Wired:

"For more than a century, bacterial cells were regarded as single-minded opportunists, little more than efficient machines for self-replication...The sole ambition of a bacterium is to produce two bacteria.

"New research suggests, however, that microbial life is much richer: highly social, intricately networked, and teeming with interaction...

"Researchers have determined that bacteria communicate using molecules....

"Microbes are able to collectively track changes in their environment, conspire with their own species, build mutually beneficial alliances with other types of bacteria, gain advantages over competitors, and communicate with their hosts – the sort of collective strategizing typically ascribed to bees, ants, and people, not to bacteria.

"Their discoveries suggest that the ability to create intricate social networks for mutual benefit was not one of the crowning flourishes in the invention of life...

"It was the first.

"'Cell-to-cell signaling' communication seems to be the rule, rather than the exception, in every domain of life."

And more, from ScienceDaily:

"There is a perception that single-celled organisms are asocial, but that is misguided. When bacteria are under stress, which is the story of their lives, they team up and form this collective called a biofilm.

"If you look at naturally occurring biofilms, they have very complicated architecture. They are like cities with channels for nutrients to go in and waste to go out.

"...scientists have shown that bacterial communication within populations can be disrupted by the invasion of cheater cells who either do not produce the communicative signal or do not bother listening and responding to signals made by other cells.

"However, bacteria tend not to cheat when dealing with their close kin…

"'We can no longer consider bacteria to be single celled entities living and dividing in isolation of each other. They can communicate with each other, preferentially direct aid towards close relatives and even cheat on each other.

"Bacterial populations are a lot more sophisticated than many people have thought.'"

Even more from ScienceDaily:

"Microbes may be smarter than we think. A new study by Princeton University researchers shows for the first time that bacteria don't just react to changes in their surroundings -- they anticipate and prepare for them. The findings, reported in Science, challenge the prevailing notion that only organisms with complex nervous systems have this ability.

"What we have found is the first evidence that bacteria can use sensed cues from their environment to infer future events,

"The two lines of investigation came together nicely to show how simple biochemical networks can perform sophisticated computational tasks,

"When a biofilm composed of hundreds of thousands of bacterial cells grows to a certain size, the protective outer edge of cells, with unrestricted access to nutrients, periodically stopped growing to allow nutrients to flow to the sheltered center of the biofilm.

"In this way, the protected bacteria in the colony center were kept alive and could survive attacks by chemicals and antibiotics.

"…oscillations in biofilm growth required long-range coordination between bacteria at the periphery and interior of the biofilm…

"…the coordination among distant cells within biofilms involves a form of electrochemical communication.

"'Just like the neurons in our brain, we found that bacteria use ion channels to communicate with each other through electrical signals. In this way, the community of bacteria within biofilms appears to function much like a microbial brain.'"

Ants self-organize into communities of aphid or fungus farmers, agronomists for millions of years before human farmers arrived on the scene.

Birds self-organize into murmurations and migrations, fish into swarming schools, bees into throbbing masses for the protection of the community of birds, fish and bees.

We are even discovering that our brains self-organize to heal themselves (sometimes) in what still seems to be miraculous ways.

Proteins in our cells self-organize as a matter of course to accomplish routine cellular maintenance and behavior. The cells self-organize into organs, and the body is an interactive complex system that operates with intent and specificity largely outside of our direct control.

Humans self-organize, of course, into families, communities, societies, political parties, sports fans, and drunken frat parties, not to mention riots and armies.

Linux as an operating system is a result of human self-organization. So is Firefox. Wikipedia. The Internet itself. Reality shows. Football matches. Games. Dependent on which football is the football you like. Language. Bad language. Culture. Bad culture. See "frat parties" and "soccer hooligans".

So not only is the universe and everything in it a product of interaction, those interactions ultimately draw the universe and everything in it into a complex, interrelated, interdependent web of relationships.

It's not just a global community. It's a cosmic community.

Community is what the universe is all about. The cosmos self-organizes into, well, interactive communities with intent and specificity according to the laws of physics. The laws of physics themselves are instruments of self-organization, an interaction between particles and forces, neither of which is actually there in the way that we think of "there" to be.

The result is "emergence". What emerges is unexpected, unpredictable, unlikely in the extreme, and maybe largely unknowable and incomprehensible. It is the cosmos, empty except for the interactions that give it birth and life and death. And reality TV shows.

And life itself, consciousness itself are likely to be the result of self-organizing emergence.

Chapter 53   Community

Reality TV shows are actually a good, which is to say, terrible example.

A bunch of strangers (emphasis on the word "strange") are gathered together by an unseen power (God, or the producers), given a bunch of rules, and then we all watch as they self-organize into a community that becomes

1) greater than the sum of its parts (note: "greater" does not necessarily equal "great")(the quality of the parts i.e. contestants sets a pretty low bar for "great"),

2) unpredictable in producing an outcome, but the process is

3) emergent. It emerges from the complex interactions of the people with each other and the rules, along with the things that happen along the way.

The point of the show is not the rules; it is to be found in the emerging interactions and resultant community, albeit dysfunctional and cannibalistic, in that the contestants are gradually consumed by one another and disappear from the show.

Reality shows are a bit like the Universe a la Big Bang, as well.

They do not exist at all, and then they do.

They come into being (though not quite as suddenly and with a large amount of entropy) out of the interactions of the creators, producers, directors, then cast and crew.

The rules don't create the show – the show creates the rules, the matter and energy emerging from the creative process.

The universe is about producing communities, whether the communities are cellular or galactic super-clusters in size. The communities do far more than the parts alone could ever do, they are produced by entropy in ways that make no common sense but are true, and are both predictable and unpredictable, giving us stars, elements, planets and life, and also advertising and TV ratings, but without giving us any specific hints about what kind of stars, life, or ratings.

 The communities emerge from the interactions and produce interactions of their own. All interactions emerge from interactions, but unpredictably.

So if God exists, it seems likely that he is an interactive God and is all about those interactions producing communities.

Faith communities, as it happens.

God is not about the rules. God is about the interactions that are communal in their very design, purpose and process.

And hence, both the joy and sorrow of religion. Religion is supposed to be an emergent community based upon the rules of the faith, and that does happen, but what always also seems to happen is that each religion becomes driven and contained by the rules rather than released by them into relationship.

It is not a freeing process but a constraining one, not an inclusive process, but an excluding one.

Religions become very much like reality TV shows, the fanatical adherence to the laws of the faith winnowing down the faithful until only the few are left.

As long as you accept the strictures, you will be accepted, loved and cherished, but fall a little, and you might just be invited off the island.

Take Christianity.

Well, first, of course, you have to take Judaism, born out of an interaction between God and man, and out of that interaction, rules emerged, simple, basic, logical reasonable rules to define that interaction. Love God. Don't kill each other, or steal from each other. And so on.

Then the Jews started trying to figure out how to finesse the rules.

So the rules got more specific, more burdensome, more restrictive, and Judaism became effectively closed to anyone who might what to join

– once again, that circumcision thing is not a good business model.

The roles switched. Judaism became less about relationship and more about rules, obsessively.

Islam is the same. Buddhism and Hinduism, the same. Jainism, ditto. Taoism, Confucianism, all the –isms.

So Jesus comes along to fix it by reducing the rules to two: Love God, love each other as you love yourself.

But then Christianity became the same. The relationship subsumed by the rules. The rules became the relationship.

It was the pathway to an interactive community.

But immediately, followers began to try to figure out what it meant to love God. What rules to we have to follow to make that happen?

Those two rules that Jesus gave us are inclusive. They do not exclude.

So we had to come up with a whole new set that excluded. I'm in, and you're out.

In a sense, it's what human communities do. They include and exclude. It's a safety thing – strangers are sometimes dangerous and threatening, so we need to be able to identify them. As well, there are things one might try to do that are dangerous to others and self, so we need to identify the danger and stay within the boundaries.

It's a belonging thing – we have a deep, deep need to belong to something greater than ourselves.

It's a meaning and purpose thing - that thing to which we belong that is greater than ourselves gives our lives meaning and purpose.

It is a conforming thing – we feel safer and more comfortable when we are with those who are most like us.

It is a stress-reducing thing – "different" is difficult, requires thought and adjustment. "Similar" is easy, no thinking required.

But it is an excluding thing, and therefore it is not completely a Jesus thing.

So the question arises – how to fix it?

How do we stop a natural and normal process of getting organized and coming up with rules of behavior to subsume and eventually emasculate religious communities?

Well.

You can dictate and control the process, dominating all the people with a authoritarian central authority figure who has been granted semi-divine status and rules with fear and threats of ostracism, maybe even torture and death as common options, plus the ever popular condemnation to hell for refusing to conform.

That's been done. Is being done. Without naming any names.

Otherwise, history seems to be full of other outcomes.

In some cases, the faith withers and fades into a marginalized status as it loses members.

In others, the members of the faith become so different in appearance and actions from the surrounding super-culture that they are ridiculed, attacked, maybe killed or arrested.

Pogroms arrive.

They may become so different that they become a cult and reject the super-culture altogether. Maybe Kool-Aid is involved.

Or evolve into an uncomfortable hybrid, a cult large enough to persist and attain a certain level of unhappy tolerance by the super-culture. Scientology springs to mind. Jehovah's Witnesses. The Moonies. (Used to know a Moony.) The LDS. In the latter case, a conscious attempt has been mind to become more like mainstream Christianity and therefore more acceptable. How sincere that attempt is remains to be seen.

But the most insidious and common is to be dissolved into the culture at large. Examples abound. You'll have to read my other books to get filled in on all of that.

Chapter 54 — All You Need Is Love Love. Love Is All You Need.

Summing up. Again.

Short and sweet.

To the point.

Don't drag it out.

Just get on with it.

OK, then.

There's nothing in the universe.

Energy, whatever the heck that is.

Matter, which is the same thing.

Space-time, which is, um, nothing except that it's flexible and is full of seething virtual particles which aren't really there except in a persistent enough way to give the illusion of actually being there.

Big Bang = Everything Came From Nothing.

Which seems like a problem if you actually think that the Everything That Came From Nothing Is Actually Something.

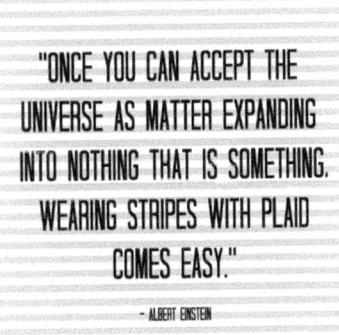

Pretty sure Albert didn't say this.

But as it turns out

The Something that is the Everything that Came from Nothing is actually Nothing.

Except for.

Interactions. a.k.a Laws of Physics.

Which really all that is there.

The problem that we all have is that we persist in thinking that things are actually there.

That they exist.

As solid things.

But solid things are made of much smaller things that aren't actually there.

Neils Bohr said that. Or something like it.

But what IS there is

Relationships.

Interactions.

That's what the universe actually is.

Interactions.

And that goes all the way from the beginning, when Big Bang needed a quantum interaction to become a Big Bang, and hence a universe with all the bits that a universe needs

to me having a crush in the 4th grade on a pretty little girl in Albuquerque New Mexico in 1962.

For example.

She wore glasses. She was hot. Even though we didn't use that word in the 4th grade in Albuquerque New Mexico in 1962.

Anyway.

The universe is all about interactions.

And the observational interactional role that a human has to play turns out to be

essential

for the existence of the universe itself.

Dang.

It really only needs one observer.

The rest of you are

extra.

Back-up.

Use only in an Emergency.

Now.

If this is all true,

and today it still is,

and if God exists,

and the evidence is pretty danged good

(For example

Big Bang. Order. Structure. Laws of physics. Everything comes from nothing. Fine tuning.)

then God stands a danged good chance of being the kind of God that is all about

Interactions.

Relationships.

Um. Love.

God is love.

Short and sweet. To the point. No dragging it out needed.

Then in order for you (I'm talking to *you* now) to do what God and the universe seem to need to make things right

love God. And love other people.

So.

Go do that.